NEUROGASTRONOMY

COLUMBIA UNIVERSITY PRESS NEW YORK

NEUROGASTRONOMY

How the Brain Creates Flavor and Why It Matters

Gordon M. Shepherd

Columbia University Press
Publishers Since 1893
New York   Chichester, West Sussex
cup.columbia.edu

Library of Congress Cataloging-in-Publication Data
Shepherd, Gordon M., 1933–
Neurogastronomy : how the brain creates flavor and why it matters /
Gordon M. Shepherd.
p. ; cm.
Includes bibliographical references and index.
ISBN 978-0-231-15910-4 (cloth : alk. paper)
ISBN 978-0-231-53031-6 (e-book)
I. Title.
[DNLM: 1. Olfactory Perception—physiology.   2. Brain—physiology.
3. Taste Perception—physiology. WV 301]
LC-classification not assigned
612.8—dc23
2011029170

c   10   9   8   7   6   5   4   3   2

for Grethe

Contents

PART III
Creating Flavor

PART IV
Why It Matters

Preface

Eating is our most common behavior. We normally do it every day through-out our lives. Scientists over the past 20 years have elucidated how our urge to eat is largely controlled by hormones, that turn on when we are hungry and off when we are full. But this hormonal control doesn't explain why we like certain foods and not others; why we may crave too much of what we like or too little of what we don't like. To address these questions, a new science of eating is emerging that focuses on food flavors. A common misconception is that the foods contain the flavors. Foods do contain the flavor *molecules*, but the *flavors* of those molecules are actually created by our brains. If we are to eat healthfully and avoid the many chronic diseases that are affected by poor diet and nutrition, it is important that we learn how the brain creates the flavors that we experience—in short, we are embarking on a new scientific endeavor that I have called neurogastronomy.

I have been led to this new field by working on how the brain creates images of smells. These findings and other studies from laboratories around the world are radically changing the common view of the sense of smell, from being one of the weakest of our senses, to being, through its role in flavor, one of the most important in our daily lives. This work is leading to a new concept of a unique human brain flavor system, perhaps the most extensive behavioral system in the brain, creating perceptions, emotions, memories, consciousness, language, and decisions, all centered on flavor. By combining brain studies with food studies, and drawing on the wisdom about flavor exchanged within families every time they eat together, neurogastronomy holds the promise of putting healthy eating on a new scientific basis. In this book, I draw on these studies to explain this new field and show that it holds benefits for everyone.

Acknowledgments

I have many to thank for helping to make this book happen.

Jean Black at Yale gave me early encouragement to write a book on smell. After I had given up the project as being too difficult, my colleague Stuart Firestein revived both me and the manuscript and put me together with Patrick Fitzgerald at Columbia University Press; they both insisted that I could do it. I am grateful to the editors at *Nature* magazine for inviting me to write an "Insight" article on smell. Christian Margot persuaded me to keep the focus of that article on retronasal smell and flavor, and that focus led directly to this book.

The theme of smell reflects my many colleagues who have worked with me over the years on how the smell pathway constructs activity patterns of smell molecules that are the basis of smell perception. In addition to being indebted to Stuart, I am grateful to Charles Greer for support at Yale, as well as former students Lewis Haberly, John Kauer, Tom Getchell, William Stewart, Kensaku Mori, Doron Lancet, Patricia Pedersen, Frank Zufall, Trese Leinders-Zufall, Wei Chen, Minghong Ma, Xavier Grosmaitre, and David Willhite, who feature in these pages, and visitors Dennis Lincoln, Burton Slotnick, and Matthias Laska.

I am grateful to Terry Acree for inviting me to be on the faculty of a workshop on flavor, through which I met Harold McGee, who has become a wonderful friend and source of knowledge and inspiration.

My wife, Grethe, and my daughter, Lisbeth, are Francophiles, and this has opened many doors. Beginning with a sabbatical in Paris in 1986, I have been privileged to interact many times with Jean-Didier Vincent, one of France's leading scientists and philosophers in the worlds of wine and culture. He and several close friends gave Grethe and me an unforgettable evening of ortolan. Through Jean-Didier I took part in a national radio program in France on wine in 2000, another stimulus along the way to this book. Through these activities I met Jean-Claude Berrouet, chief wine taster of the house of Petrus, who provided me with an unforgettable personal tasting of 10 different Petrus wines of different ages, an opportunity to test the idea of images of wines. Pierre-Marie Lledo has been a gracious host on many visits to Paris to pursue common interests in experiments on olfaction in the laboratory and experiments on tasting wines outside.

For my training as a neuroscientist and for laying the foundation for this book, I was fortunate to begin with studies at Oxford University under Charles Phillips and Thomas Powell. Tom was in the Department of Anatomy chaired by Wilfrid le Gros Clark, who was also a leading anthropologist. Through this connection I began my interest in the anthropology of the human sense of smell. In taking up this interest more recently, I was greatly aided by Richard Wrangham and Dan Lieberman at Harvard, starting with a seminar to the Department of Anthropology and continuing with extended discussions with Dan on the evolution of the human head, including the retronasal passageway. After Oxford, Wilfrid Rall at the National Institutes of Health has been an extraordinary friend and mentor, together with colleagues Thomas Reese and Milton Brightman. Later, Frank Sharp and Ed Evarts at the National Institutes of Health led our way to discovering the brain activity patterns representing smell molecules.

My education in the rich world of flavor and the brain has benefited from many colleagues, especially Linda Bartoshuk, Valerie Duffy, Terry Acree, Christian Margot, Avery Gilbert, Bruce Halpern, Marcia Pelchat, Richard Doty, Anthony Sclafani, and Ed Rolls. I have had valuable advice in putting together this book from Valerie Duffy, Justus Verhagen, Dana Small, Daeyeol Lee, and Gary Beauchamp, who have read chapters and offered criticism in their areas of expertise. Harold McGee and Donald Wilson have provided valuable feedback on the entire manuscript. Any residual errors are mine.

For expert guidance in seeing the manuscript through production, I'm first grateful to Wendolyn Hill for transforming my scientific slides into the simple and elegant illustrations for this book. Bridget Flannery-Connor has provided expert oversight, Pamela Nelson a sure hand in guiding the production, my editor Patrick Fitzgerald his always unwavering support from beginning to end, and Stuart Firestein with his always wise counsel. Grethe and my family have borne with me throughout, for which I can only say, "Tak! Mange tak!"

The research on which this book is based has been carried out with support from the Chemical Senses Program of the National Institute for Deafness and Other Communication Disorders, under the National Institutes of Health in Bethesda, Maryland. My main grant has been funded continuously for 42 years. I have also received support from the National Science Foundation, the National Aeronautics and Space Administration, the National Institute on Aging, the National Institute for Drug Abuse, and the Naval Medical Research Laboratory. All of this work has been funded through periodic competitive reviews. I am deeply grateful to the study section committees for their evaluations, and the councils of the institutes for the funds. Our projects have been for animal studies of microcircuits in the smell pathway; the fact that they revealed insights that led to new ways to think about human flavor is one of the great rewards in science, of having the freedom to follow basic research wherever it may lead.

NEUROGASTRONOMY

INTRODUCTION

Retronasal Smell and the New Age of Flavor

The origins of this book are the cooked meal around the family dinner table that my wife, Grethe, and I have shared at the end of each day beginning when we were students together.

Her interests run to food (as a gourmet cook), books (as a reference librarian and an avid reader), flowers and gardens (at our homes in the United States and Denmark), travel, friends, opera, and keeping tabs on our growing family. My life has been in the laboratory, studying the part of the brain responsible for the sense of smell. Through the years—from England to Washington, D.C., Stockholm to New Haven, Philadelphia to Paris, our home in Connecticut to our summer house in Denmark—through student life and family life, that shared meal has been a constant bond. A sip of sherry or wine or fruit juice, with a nibble, to prepare the digestion for a traditional meal, with meat, vegetables and salad, a glass of wine, dessert or fruit, and tea (coffee is for the morning), that varies every day. Always with Grethe's elegant touches drawn from her native Denmark and the places we've lived and visited, and from burgeoning piles of recipes clipped from newspapers and magazines from around the world. We treasure the time to enjoy the flavors of the food while reflecting on the events of the day.

This daily routine would have gone on without reflection except that in 1986, *National Geographic* prepared its historic article "The Intimate Sense of Smell," the first comprehensive public overview of this neglected sense. As an olfactory scientist, I was interviewed by the author. You may have seen photographs from this article, the most notorious being of a row of young men stripped to the waist with upraised arms while a row of pleasant middle-aged women in white lab coats have their noses buried in the boys' armpits, testing the effectiveness of the latest underarm deodorant.

The article started by noting the common idea that smell seemed to have faded in importance for humans when our ancestors started walking upright, relying on the visual sense: "But to Gordon Shepherd, a Yale University neuroscientist, people vastly underrate their sense of smell. 'We think our lives are dominated by our visual sense,' he said, 'but the closer you get to dinner, the more you realize how much your real pleasure in life is tied to smell. It taps into all our emotions. It sets the patterns of behavior, makes life pleasant and disgusting, as well as nutritious.' "

When the author read back to me this quotation that he was planning to use, I protested that I must have phrased it more elegantly, but he insisted that it was what I had said so that is the way it stayed. Anyway, my main point was to oppose the common wisdom that the sense of smell has become weak and insignificant in humans (unless you are a perfumer) by remembering what the aromas and flavors of those daily dinnertimes had meant to me.

Later, it began to dawn on me that I must be pretty dense to be working all my life on the physiology of the sense of smell without trying to figure out how it applied to enjoying my evening meal. It was time for me as a neuroscientist to think about how those smells from the food in my mouth were able to reach the sensory cells deep inside my nose, and how those smells were merged with other sensations to produce flavor.

Searching for those answers sent me on an odyssey that has been fascinating every step of the way. It has introduced me to many investigators working in areas unknown to one another, to the mainstream of neuroscientists, and to the general public. I learned about *food scientists*, who study how food is chewed in the mouth and how it is swallowed. *Physiologists* study how the smells are carried to the sensory cells in the nose by our breathing in, as when inhaling the aroma of a cup of coffee, and by breathing out, as when we chew and swallow our food. *Psychologists* study how smell is combined with taste and the other senses to produce what is called *taste* but is really *flavor*, one of the most complex of human sensations. *Cognitive neuroscientists* use brain imaging to show how flavor arises from activity at the highest cognitive levels of our brains. *Neuropharmacologists* study how the regions of the brain that are activated by cravings for food involve some of the same regions that are activated by cravings for tobacco, alcohol, and drugs of abuse. *Biochemists* identify the hormones that circulate in the blood, connecting our bodies with our brains to signal to start eating when we are hungry and stop when we are full. *Anthropologists* speculate that cooked food was a major driving force in human evolution. *Molecular biologists* have discovered that the sensory receptors for smell form the largest gene family in the genome, and they are studying how the molecules give rise to our perceptions of different smells. And we are good smellers: *behavioral psychologists* find that monkeys as well as humans have much more sensitive senses of smell than previously recognized.

All these investigators, most of them without knowing one another's fields, are building the new science of flavor. In addition, food critics have been pushing and pulling us into a new age of food appreciation. They include Harold McGee, whose book *On Food and Cooking: The Science and Lore of the Kitchen* has educated thousands and entertained them with the ways in which foods give rise to flavors. An important step on this path was taken by Nicholas Kurti and Hervé This, who became fascinated with the "physical and chemical aspects of cooking" and began to hold workshops on their new passion. They helped a new field to emerge, defined in This's book *Molecular Gastronomy: Exploring the Science of Flavor* as the science dealing "with culinary transformations and the sensory phenomena associated with eating."

Among all this work, several things stood out to me:

1. When we sense the flavor of the food in our mouths, it is not by sniffing in, which we usually associate with smelling something like an aroma, but by breathing out, when we send little puffs of smell from our food and drink out the back of our mouths and backward up through our nasal passages as we chew and swallow. This back door approach is called *retronasal smell* (*retro = backward*); we can also call it *mouth-smell*. It contrasts with *orthonasal smell* (*ortho = forward*), which is what we call the common sniffing-smell.

2. As delivered by the retronasal route, *smell dominates flavor*. We often characterize our food in terms of how it "tastes," but the sense of taste as properly defined consists of sensitivity only to sweet, salt, sour,

bitter, and umami. What we call the *taste* of our food beyond these simple sensations should be called *flavor* and is mostly due to retronasal smell. Retronasal smell is the new frontier for studies of smell and flavor and is the dominant factor around which to build a field to study how the brain creates our sense of flavor. Simple tastes are hardwired from birth, whereas retronasal smells are learned and thus open to individual differences. They account, therefore, for the vast variety of cuisines in the world—and for why everyone who walks into a McDonald's wants a different combination of burger, nuggets, fries, salad, dressing, pastry, and cola. Vive la différence!

3. Our own studies have shown that sniffing in a smell gives rise to a spatial pattern of activity in the brain. These patterns function as *images of smell*, with different images for different smells, much as different faces form different images in our visual system. Human brains are very good at recognizing faces, which can be thought of as a highly developed form of pattern recognition. From our studies we think that the same ability occurs with the patterns laid down by smells—that is, the ability to recognize many different patterns representing as many different smells.

4. Humans have *big brains*. Although our sensory apparatus may not possess as many receptor molecules or receptor cells as that of other mammals, this fact should not prevent us from having a strong sense of smell. Humans do not have ears or eyes as sensitive as those of some of our animal brethren, either, but this did not prevent our ancestors from developing advanced abilities such as language, because we had large brains that could carry out complex processing not possible with smaller brains. During my odyssey through the different flavor fields mentioned, I wrote an article in 2004 to put forward the new hypothesis of the high abilities of the human sense of smell, which I called "The Human Sense of Smell: Are We Better Than We Think?" My colleague Avery Gilbert, who himself has written an excellent book on smell, *What the Nose Knows: The Science of Scent in Everyday Life*, sent me an e-mail with the comment that my title was not quite right. He said it should have been "The Human Sense of Smell: It Is Better *Because* We Think." Precisely. The big brain is especially important for flavor. A key premise of this book is that *humans have a much more highly developed sense of flavor because of the complex processing that occurs in the large human brain*. It is this high level of processing—including systems for memory, emotion, higher

cognitive processing, and especially language—that give us what I call our unique *human brain flavor system*.

In summary, up to now the focus of food science has been on relating the composition of the food to the perceptions of the flavors. Some of the explanations have begun to reach into the mechanisms of the brain. What is now needed is to begin with the brain, and show not only how it receives the sensory stimuli, but how in doing so *the brain actively creates the sensation of flavor*. It is important to realize that flavor does not reside in a flavorful food any more than color resides in colorful object. Color arises as differences in wavelengths of light given off by an object; our brains transform those wavelengths into color to give it meaning for our behavior. Similarly, the smells that dominate the sense of flavor arise as differences between molecules; our brains represent those differences as patterns and combine them with tastes and other senses to create smells and flavors that have meaning for our perceptions of food.

Understanding how the brain does this constitutes a new science of flavor. In 2006, after the article on how the human sense of smell is better than we think, I reviewed the different fields of study contributing to the new understanding of the brain and flavor for a special issue of the journal *Nature* called "Nature Insight." I needed a term to apply to all these fields coming together in the study of flavor, and the word *neurogastronomy* popped out, so that is what I will call it. It builds on the old term *gastronomy*—first conceived by the ancient Greeks to apply to eating well and living well, and popularized as such in France in the early nineteenth century—and adds *neuro* to represent the brain. Other fields start with the food, and ask how it stimulates the senses. Neurogastronomy starts with the brain, and asks how it creates the sensations of the food.

This new field is timely. In the past few years, the role of smell in flavor, and the way in which it arises internally from the back of the mouth, have received increasing research interest. However, this role still remains hidden to most people. I am a little shocked when I talk about my newfound passion for retronasal smell and its role in flavor and some of my science colleagues express surprise on hearing that we sense flavor when we breathe out, and that flavor is not all due to taste. How can we hold informed opinions on food and nutrition, gourmet eating, fast food, and obesity if the role of smell is not recognized and understood for its dominant function in the perception of flavor, if flavor itself is not recognized for the dominant role it plays in our daily lives, and if the brain regions involved in cravings for food are not recognized to be the brain regions involved in cravings for drugs of abuse? It is time for a new appreciation of how the brain creates smell and flavor for the twenty-first century.

This book lays out some key elements of the new science.

Chapter 1 gives background on the new ways of thinking about smell as a dominant sense for humans because of its central role in producing flavor. Chapter 2 shows how a dog's sense of smell is beautifully adapted for tracking while sniffing in, whereas a human's is equally beautifully adapted for flavor while breathing out. Chapter 3 warns about how the mouth fools the brain into thinking it is producing all the flavor, with experiments you can do to prove it to yourself. Chapter 4 provides an introduction to the molecules in food that give them the flavor that the brain creates.

Chapters 5 through 12 describe the smell pathway in all its glory as the main actor in the flavor system, focusing on the new evidence that the information in the smell molecules is represented in the brain as a smell image, that this occurs through retronasal smell, and that it is subject to a basic neural operation called *lateral inhibition* to enhance discrimination. We take a detour along the way to show how a smell image draws on our understanding of how complex images in vision work, such as in recognizing faces. Chapters 13 to 16 then show how the other senses—taste, touch, vision, and hearing—play their own important roles in the total sensation of flavor. Chapter 17 brings in all the muscles and movements that bring about the sensing of what is in the mouth and how it leads to retronasal smell.

All this complex machinery is summarized in chapter 18 as the "human brain flavor system," which is at the heart of neurogastronomy. This system is closely interconnected with regions of the brain that produce emotion by creating "images of desire," as discussed in chapter 19. Chapter 20, through the iconic tale of Proust and his madeleine cookie, highlights the power of flavor to evoke memories. Chapter 21 shows how fast foods capture the human brain flavor system with excessive activation, and how this leads to poor nutrition and obesity. Why do we choose the foods we do? Chapter 22 introduces the reader to the new field of neuro-economics. As applied to the brain, it seeks to explore how the flavor system makes decisions that lead us to eat healthy or unhealthy foods. There is increasing evidence that loss of control over those decisions involves the same parts of the brain involved in drug addiction. All of this activity occurs in brain microcircuits that are plastic; that is, our brains are changed by their own activity, as explained in chapter 23. Chapter 24 addresses the fascinating and frustrating link between smell and language, suggesting that smell shares with vision and hearing the challenge of using words to characterize complex images. Chapter 25 explains how perception of smell and flavor gives us new insights into the neural basis of consciousness. The dominant role of retronasal smell in our daily eating behavior implies that it must have played an important role in the evolution of the human brain, a new hypothesis that is discussed in chapter 26.

Why all this matters for public policy is discussed in chapter 27, which addresses key questions of how the brain begins before birth to create flavor preferences that can last a lifetime; how infants and children are especially attracted to both healthy and unhealthy flavors; what the relation is between flavor and nutrition; the critical role that flavor plays in diseases such as obesity and hypertension; and finally, how to use flavor to make better lives for the ill and the elderly.

PART I

Noses and Smells

CHAPTER ONE

The Revolution in Smell and Flavor

I met a man from Denmark who had been to Disney World in Florida for a week with his friends.

"How did it go?" I asked. "Did you enjoy American culture?"

"Oh yes," he said. "We had a great time."

My wife is Danish, and I know how much a Dane loves food, so I kidded him by asking, "Did you enjoy eating hamburgers and hot dogs every day?"

"We never touched them," he said.

"What in the world did you live off?" I asked.

"We brought with us a week's supply of *rugbrod* [Danish pumpernickel] and Danish salami and Danish cheese. We even had some smoked eel."

A week's supply of that stuff would have weighed a ton. I know; I've lugged suitcases of *rugbrod* across the Atlantic myself when returning from trips to the Old Country. The deep dark aroma and bouncy chewiness of Danish *rugbrod* is unique. So is a bite of salty, meaty Danish salami, as well as the familiar flavor of Havarti cheese with dill. For me they're a part of my adopted identity when I'm in Denmark. But for my Danish friends they were their native identity.

This story recognizes the old saying "Patriotism is a longing for the food of our homeland." It expresses a loyalty to our home country based on the flavors of the food we were brought up on. When athletes in the Olympics play for the glory of their country, they also in a way play for the flavors of their country's foods. Even in our age of globalization, when our daily fare may include dishes from other lands—think sushi in Los Angeles, pasta in New York, or "McDo" in Paris—the particular combinations we learn while growing up are part of our national identity. On a vacation trip in a foreign country, how strong is our desire, after tasting a few meals of the local food, for a burger, or fish and chips, or sushi, or lasagna, or . . . you fill in the blank. I like soya flavor, but on a trip to Japan, after days of around-the-clock soya, I yearned for a familiar flavor, so Grethe bought me small cartons of cold cereal and milk that I could have for my traditional breakfast.

Given this importance of the flavors we learn to like, it seems to me remarkable, and unfortunate, that most people are unaware that the flavors are due mostly to the sense of smell and that they arise largely from smells we detect when we are breathing out with food in our mouths. Few people know how modern research shows that an odor sets up patterns of activity—"smell images" in our brains—that are the main basis for our perception of flavors. These smell images are hidden factors that determine most of the pleasure we get from eating, and they share the blame for the problems we incur when eating foods that are not good for us. If we can understand better the central role that smell plays, we can understand better how to reduce the problems and increase the joy.

To appreciate the importance of retronasal smell in our lives, let's step back and look at how far we've come in changing ideas that go back to the ancient Greeks.

Most people regard the sense of smell as not very important. The legacy of thinking of smell in this way began with Aristotle. In discussing the senses in *De Anima* (*On the Soul*), he observed that "our sense of smell is inferior to that of all other living creatures, and also inferior to all the other senses we possess."

If you're laboring under this misapprehension 2,500 years later, you're not the only one. Like almost everyone else, you probably enjoy pleasant scents, such as attractive perfumes, fragrant flowers, and a steak barbecuing on the grill, and you do not like unpleasant smells, such as body or bathroom odors and polluted air. And that is about the limit of how much regard you give to smell.

These impressions seem minor compared with the important roles played by our sight and hearing. Against these, smell can seem trivial, although not if you suddenly lose your sense of smell because of an accident or infection. In her book *Remembering Smell: A Memoir of Losing—and Discovering—the Primal Sense*, Bonnie Blodgett has described the devastating loss of flavor that may occur. Most of that loss is due to retronasal smell. Still, if forced to give up one of the senses, who would choose a life without vision or hearing rather than smell? Sight and sound are obviously the essential senses for normal living and use of language.

But this is putting the question in the wrong way. What are the factors that shape our daily behavior? Which inputs to our brains are the motivating forces that determine the quality of our daily lives and that influence the decisions we make about health, diet, mates, and social relations? If we stick with only our obvious sensations, we miss the deeper factors. Among them, smell plays powerful but hidden roles.

As discussed in the Introduction, the new evidence regarding these roles for smell comes from work in many fields. Taken together, all these studies reflect how the sense of smell and associated flavor engage an astounding extent of the human brain. This work is not only intriguing to the general public, but involves profound insights into our biological nature. Many research workers are realizing that the sense of smell is ripe for investigation as one of the most exciting frontiers in the brain. They are intrigued that this system may hold the key to unlocking many of the secrets of our body. This was already realized decades ago by the physician and essayist Lewis Thomas: "I should think we might fairly gauge the future of biological science, centuries ahead, by estimating the time it will take to reach a complete, comprehensive understanding of odor. It may not seem a profound enough problem to dominate all the life sciences, but it contains, piece by piece all the mysteries."

One of those mysteries is how the brain uses smell to create flavor.

Current studies are already revealing capabilities of human smell that go far beyond the traditional view. Rather than being weak and vestigial, human smell appears to be quite powerful. Some have even suggested that humans and their primate relatives are "supersmellers" among animals. It is time, therefore, for a new appreciation of this much maligned and neglected sense. The aim of this book is to show how the real power of human smell lies in its key role in human flavor.

Flavor as a Force in Human History

Despite the dim view that Aristotle took of the sense of smell, this sense, through its dominant role in flavor, had its revenge by shaping the course of world empires during human history. Eric Schlosser, in *Fast Food Nation: The Dark Side of the All-American Meal,* was very clear about the dominant role of smell in flavor: "'[F]lavor' is primarily the smell of gases being released by the chemicals you've just put in your mouth." And, he was also clear about the spell that flavor exerts over humans:

> The human craving for flavor has been a largely unacknowledged and unexamined force in history. Royal empires have been built, unexplored lands have been traversed, great religions and philosophies have been forever changed by the spice trade. In 1492 Columbus set sail to find seasoning. Today the influence of flavor in the world marketplace is no less decisive. The rise and fall of corporate empires—of soft drink companies, snack food companies, and fast food chains—is frequently determined by how their products taste.

Thus not only is flavor not recognized for the force it has been in human history, but smell has not been recognized for the dominant part it played. This despite the fact that flavorful herbs and spices were essential to Roman cuisine more than 2,000 years ago. As the quotation indicates, the quest for new flavors and aromas was one of the driving forces behind long-distance travel and the opening of trade routes. It is not widely appreciated that a millennium before Marco Polo "discovered" China, the spice trade of the Roman Empire stretched from the Mediterranean to China and the South Pacific, delivering to southern European tables delightful flavors and aromas to enliven their daily fare. The voyage of Christopher Columbus was not embarked on to prove that the world was round, but to find a more direct route to the sources of spices because of their flavors.

The voyagers who followed in the sixteenth and seventeenth centuries were seeking not just gold, but also control of spice trade and production. The atrocities committed over several centuries in the desperate attempts to control those sources are among the worst of any time. On the positive side, the case has been made that the organization of the great sea lanes for shipping tea and spices by fast clippers from Indonesia in the eighteenth and nineteenth centuries laid the basis for the British Empire, followed by the emergence of world powers and their economies in the nineteenth and twentieth centuries. We have inherited and adapted those global trade routes, from controlling the trade in tea and spices to controlling the sources of oil and spreading global capitalism in the past century.

There has thus been a contrast between the received wisdom that smell and flavor are of minor importance and the realities of the high value that humans put on them in their daily lives, as well as of the consequent economic forces that have driven human societies.

This book will provide new insights into why the smell and flavor of the foods we eat have such high value in human affairs. There have been some key steps along the way to our present understanding.

The First "Physiology of Flavor"

It was not until Jean Anthelme Brillat-Savarin that smell began to be appreciated, particularly for its role in taste, equivalent to *flavor*. Born in 1755, Brillat-Savarin was a lawyer and then a mayor. During the French Revolution, he was forced to flee France, spending two years in the United States. He then returned after the defeat of the Jacobins in 1796 to become a judge under Napoleon. His new position gave him plenty of free time to spend on writing about his main passion, eating well, otherwise known as *being a gourmand*. Shortly before his death in 1826, he published his reflections as a book whose full title is *Physiologie du goût; Ou, meditations de gastronomie transcendante: ouvrage théorique, historique et a l'ordre du jour*, Dédiée aux gastronomes parisiens, par un Professeur, Membre de Sociétés Litteraires et Savants (*The Physiology of*

Taste; Or, Meditations on Transcendental Gastronomy: Theoretical, Historical, and Practical Work, Dedicated to Parisian Gastronomes, by a Professor, Member of Literary and Scientific Societies).

As implied by the elaborate title, the book was not strictly a scientific treatise. However, in its thoroughness in considering from many angles the pleasures of eating, and in one delightful anecdote after another illustrating how human society is dependent on the social interactions that take place while eating meals together, Brillat-Savarin's book became a classic. Part of this was due to his way with words; from him came such famous observations (in M. F. K. Fisher's well-known translation) as

Tell me what you eat, and I will tell you what you are.

The discovery of a new dish does more for human happiness than the discovery of a new star.

Of most interest for our purpose was his passion for revealing the physiological and psychological processes that are responsible for the perception of taste. The *taste* in his title refers not specifically to the sense of taste, but to the combined perception of taste and smell, which we refer to as flavor. Brillat-Savarin acknowledged the dominant role of smell in taste and flavor:

I must concede all rights to the sense of smell, and must recognize the important services which it renders to us in our appreciation of tastes; for, among the authors whose books I have read I have found not one who seems to me to have paid it full and complete justice.

For myself, I am not only convinced that there is no full act of tasting without the participation of the sense of smell, but I am also tempted to believe that smell and taste form a single sense.

Brillat-Savarin thus identified clearly the important role of smell in taste, but unfortunately didn't differentiate clearly between taste as a single sense and "taste" as a combined sense of smell and taste. That is why we will call the combined sense "flavor." The close relation between smell and taste has obscured the dominant role of smell. A main aim of this book is to disentangle them.

Brillat-Savarin recognized that smell's contribution to taste could come only from the smell arising at the back of the mouth and being swept into the nasal chamber by what we now call the *retronasal route*. He did not specify this route, however, or that it could be activated only by breathing out. He did declare in a colorful phrase that the back of the throat was the "chimney" of taste.

Recognition that smell's contribution to flavor came from retronasal smell and breathing out was slow in coming. In his book *What the Nose Knows*, Avery Gilbert mentions Henry T. Finck, an American philosopher who in 1886 published the essay "The Gastronomic Value of Odours" in the magazine *Contemporary Review*, in which he described how swallowing pushes the aromas from the food in our mouths into the air in the back of our throats and how exhalation carries it through the nasal chambers as "our second way of smelling." However, most interest in smell through the nineteenth and into the twentieth century was focused on trying to break down inhaled odors into a few basic categories, in analogy with the basic colors in vision. Odors, however, have been too numerous and difficult to categorize for this to be a practical venture.

The role of retronasal smell in flavor was finally put on the map by Paul Rozin, a psychologist at the University of Pennsylvania, in an article in 1982. As he phrased it, we need to recognize that smell is not a single sense but rather a dual sense, comprising orthonasal (breathing in) and retronasal (breathing out) senses. He devised experiments to show that the perception of the same odor is actually different depending on which sense is being used. Subjects trained to recognize smells by sniffing them had difficulty recognizing them when they were introduced at the back of the mouth.

Orthonasal smell is the one we commonly think of, and for good reason. It mediates a tremendous range of stimuli, evoking our sensations of the aromas of food, especially cooking food; the bouquets of wines; the fragrances of flowers; the scents of perfumes; the mysteries of incense. It also mediates the social odors: sweet scents of a loved one's breath; body excretions in sweat; volatile compounds in urine and feces; pheromone-type molecules that send signals about gender, puberty, territory, and aggression. And there are alarm signals such as for fire or gas. With these functions, all of them obvious when we detect them during normal breathing or when sniffing the air, no wonder orthonasal smell has been the main type of smell recognized over the centuries.

Retronasal smell contrasts in so many ways with orthonasal smell that it can truly be considered a separate type of smell. To begin with, it arises from inside the mouth; it is the only distance sense that arises inside the body. Second, because it arises inside the mouth, it is always accompanied by stimulation of other senses inside the mouth, mainly taste and touch. Also, because the food is inside the mouth, retronasal smell always is accompanied by movements of the tongue, the jaw muscles, and the cheeks; it is an active sense by comparison with external smell and the other distance senses, vision and hearing. Chewing releases the smell molecules from the food to be carried in the smell-laden air to the nasal chambers to stimulate the smell receptors. And astonishingly, the sense of flavor produced is a mirage; it appears to come from the mouth, where the food is located, but the smell part, of course, arises from the smell pathway. No wonder it has taken so long to begin to realize what an amazing sense retronasal smell is.

Let us take a closer look at how it works.

CHAPTER TWO

Dogs, Humans, and Retronasal Smell

To appreciate how humans are adapted for retronasal smell, it is useful to compare us with one of the acknowledged champions of smell: man's best friend, the dog. To compare the two, it is necessary to understand how each is engineered to serve its functions best. It will be no surprise that the dog's nose is an engineer's dream. But what about the human's? Are we as poor as Aristotle believed? The take-home message is that dogs are adapted primarily for sniffing in smells of the environment, whereas humans are adapted primarily for sensing smell as the main feature of flavor. Thus the dog's nose is engineered mainly for orthonasal smell, and the human nose is engineered mainly for retronasal smell. This is how the human nose fulfills its role as the key player in neurogastronomy.

Fluid Dynamics of Inspired and Expired Odorized Air

Gary Settles is professor of mechanical engineering at Pennsylvania State University. He specializes in the field of fluid dynamics, which concerns itself with the way that air or water act when passing over wings or forced through tubes. This field involves not only understanding and designing efficient wings for airplanes, nozzles for jet blowers, and ventilators for homes, but also how to draw air into devices used to detect trace substances such as explosives that might be present in cargo in container ships and in narcotics in suitcases. The latter interest led him to study how nature does it through the snout of a dog. "The Aerodynamics of Canine Olfaction for Unexploded Ordnance Detection" is the title of one of his recent research projects.

A while ago I attended a lecture in which Settles explained the fluid dynamics of the dog's snout. He began by pointing out that studies of mitochondrial deoxyribonucleic acid (DNA) from fossils support the notion that dog-like animals—including the wolf, fox, raccoon, bear, weasel, and jackal—arose around 50 million years ago in the mammalian line. This is about the time that the early monkeys were emerging in the primate line. Like other mammals, the dog-like animals moved air in and out through their noses by means of the bellows-like action of the respiratory muscles on the lungs. An exquisite coordination between smelling and breathing was thus required.

Breathing in to sample the scent in the air is what everyone knows as *sniffing*, and the dog is beautifully designed for this function. The design begins with the nostrils, technically called the *nares*. If you look directly at your dog's snout, you will see that its nostril opening has a peculiar shape, with a central round opening encircled by membranes called *alar folds* and a curved slit to the side; it has been likened to the form of a comma lying on its side. When a dog sniffs the ground, it draws in air through the central opening by muscles of the alar folds that enlarge the opening, but it breathes out by contracting other muscles that direct the outflow through the slits to the side. The singular advantage of this arrangement is that the air breathed out does not interfere with the odorized air that is being breathed in. This is particularly important as the tip of the nose gets closer to the ground. Here is how Settles describes it:

> The expired air jets . . . are vectored by the shape of the "nozzle" formed by the alar fold and the flared nostril wings. . . . Thus the external naris acts as a variable-geometry flow diverter. This has three advantages: (1) it avoids distributing the scent source by expiring back toward it; (2) it stirs up particles that may be subsequently inspired and sensed as part of the olfactory process; and (3) it entrains the surrounding air into the vectored expired jets . . . thus creating an air current toward the naris from points rostral [in front of] to it. This "ejector effect" . . . is an aid to olfaction; "jet-assisted olfaction," in other words.

It can be shown that when a dog sniffs the surrounding air, the sniff draws in air from a sphere about 4 inches (10 cm) around the opening of the nares. This is called the *reach* of the nose, for detecting odor molecules at low concentration in the surrounding air. When a dog senses the air, it changes to long inspirations, with its mouth open, in order to draw air over the olfactory membranes inside the nose slowly for careful detection. This is the way smell is used by dogs that are "pointing" after detecting the scent of a prey.

For detecting odors on the ground, the dog wants to decrease the reach distance as much as possible, which is why it has its nose virtually touching the ground when it is keen on following a scent there. The shorter distance makes the lines of air flow denser, enhancing the concentration of the inspired odor molecules. Settles has carefully charted the lines of air flow, converging on the nares inlets in the center and diverging from the slits to the side; it is a beautifully designed system for tracking scents. It enhances the high acuity of smell the way that the central fovea of the eye enhances high acuity of vision.

Settles goes further to analyze the way a dog uses this system in concert with its vision to inspect a new sense source. The dog first gets its nostrils down on the ground and sniffs its way to the scent source, following the increasing concentration until it reaches a maximum and begins to decline. It continues to sniff over and around the source; this action also enables it to inspect the source visually. The expired air plays its own role, with the lateral jets disturbing the source area enough to raise small clouds of particles that carry more odorous substances to the nares for inhaling.

Note that the dog's nose is as much a motor as a sensory organ, carrying out several motor functions: (1) acting on the environment to dislodge odors from their carriers, (2) adjusting the proximity of the nares to enhance the concentration of odors from odor sources, and (3) separating the inspiration from the expiration of odorous air to achieve the most effective balance.

The kinds of odor that a dog smells are very much a reflection of the position of this detector apparatus. In the dog, as in most terrestrial animals, standing on four legs means that the head is not far from the ground, so putting the nostrils to the ground is a natural movement, even in larger animals. The four legs also mean that the head and the hips are about level with each other. This has immediate consequences for social interactions between members of the species, because it means that when two individuals greet each other their odor detectors are not only at the same level for sniffing each other's head ends but also at the same level of each other's rear ends. Thus it is natural for dogs to leave olfactory calling cards at both ends. The same principles of maximizing odor concentration by minimizing distance apply, accounting for the intimate interactions of dogs greeting each other. It is interesting that by this means dogs acquaint themselves not only with mouth odors, signaling what has been consumed, but also with fecal and urine odors, signaling what has been digested and gone out. Through these odors they also learn of hormonal status and sexual receptivity.

So far I have discussed only the nostrils. What happens inside the snout?

Inside the Snout

Nearly all animals have some kind of a snout extending from between the eyes, containing the mouth and a cavity lined with the sensory cells for smell. In fish, frogs, and reptiles this is a simple cavity, with the odor molecules coming in through the water in the case of fish and both water and air in the case of amphibians and reptiles, directly onto the sensory cells. A long nerve tract arising from the receptor cells connects the sac to the olfactory bulb that connects to the brain. By the way, the long snout of the alligator is not long because of a long nasal cavity inside, but because the upper jaw carries a long row of teeth for crunching its prey.

When mammals arose over 200 million years ago, they are believed to have been small animals, like present-day mice or rats. Their snouts were one of their most essential adaptations to terrestrial life. In contrast to those of alligators, these were true snouts, containing an extended nasal cavity all the way out to the tip.

In a modern mammal such as the dog, the olfactory receptors are confined to the back of the nasal cavity, usually lining a series of bony convolutions in order to increase the extent of the sensory sheet (figure 2.1). In between the nostrils and the receptors is a remarkable additional organ, a kind of a cartridge even more highly convoluted and covered by

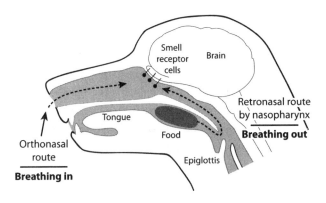

FIGURE 2.1 The head of a dog
The arrows show the pathways for sniffing smells by the orthonasal route and for sensing smells from the mouth by the retronasal route. Note the long length of the retronasal route from the mouth to the smell receptor cells.

respiratory membrane. In technical language, these convolutions are formed by the ethmo- and maxillo-turbinal bones.

Victor Negus, a British specialist in ear, nose, and throat diseases, spent many years writing what is now the classical treatise on the structure of the snout in different animals in order to compare and contrast it with the human nose. He showed that the cartridge serves as a kind of air filter or air conditioning system, with three functions: warming, moisturizing, and cleaning. *Warming* the inhaled air brings it into equilibrium with the temperature of the air in the respiratory tract. *Moisturizing* the air brings it into equilibrium with the moist air in the respiratory tract. *Cleaning* the air removes inhaled materials, such as bacteria, smoke, and particulate matter. Negus examined many different kinds of mammals and found the filtering apparatus to be present to a greater or lesser extent in all of them, with one notable exception: primates, including humans.

Just as the muscles of the nostril manipulate the inhalation of air, so are they coordinated to direct the air streams into the snout. One direction is through the middle of the air-conditioning cartridge during quiet breathing. The other is a redirection more to the side of the cartridge so that air more quickly reaches the olfactory sensory sheet, called the *olfactory epithelium*, at the back of the nasal cavity, where active sensing occurs.

When a dog is actively sniffing an object, the rate of sniffing increases dramatically, from one or two respirations a second to as many as six to

23

eight a second (small rodents like mice and rats can go even higher, to ten to twelve a second). This is much faster than a human can sniff; you can test yourself and see that the limit is about four a second.

In contrast to these adaptations for enhancing orthonasal smell, the diagram shows that the pathway for retronasal smell, from the back of the mouth through the nasopharynx to the nasal cavity, is long and relatively narrow. The dog thus appears to be preferentially adapted for orthonasal smell. What about humans?

Evolution of the Human Nose

The evolution of humans involved lifting away from the noxious ground environment as they adopted a bipedal posture. This reduced exposure of sensory cells to infections. The complicated air-cleaning apparatus thus came under decreased adaptive pressure, reducing the loss of absorbed odor molecules. The large extent of olfactory receptor cell epithelium and abundant numbers of olfactory receptors present in most mammals would have come under reduced adaptive pressure and were accordingly reduced in proportion. The result was the elimination of a snout and nearly all the air cleaning apparatus, leaving our relatively modest outward nose and inside nasal cavity.

Although this explanation seems logical, arguments from evolution are notoriously speculative. Another perspective on the evolution of the human nose is that it had less to do with the reduction of the sense of smell and more to do with the reduction of the maxilla and mandible as primates and early humans adopted a diet with less roughage.

These developments meant that the snout could be reduced in dimensions and complexity without compromising the absolute amounts of odorized air reaching the olfactory epithelium. A widely accepted scenario is that as the snout lessened in size it allowed the eyes to come forward and lie closer together to promote more effective stereoscopic vision. This supposedly allowed human evolution to be dominated by vision at the expense of smell, providing the popular explanation for how we ended up with a weak sense of smell. However, the recognition of the importance of retronasal smell for humans makes this argument irrelevant. In our new view, the reduction in orthonasal olfaction is not the key; the key is, rather, the relation of the nasal cavity to the back of the mouth, which served to enhance retronasal smell in humans.

Comparing the dog and the human shows how dramatically they differ in this respect. The dog is characterized by its elaborate cleaning apparatus for orthonasal smell, and by the long tube, the nasopharynx, connecting its nasal cavity to its pharynx at the back of the mouth for retronasal smell. The human, by contrast, is characterized by the short distance for orthonasal smell and, most importantly, the short nasopharynx for retronasal smell as well (figure 2.2). The latter is the key passageway for enabling odors released from food in the mouth to reach the smell receptors in the nasal cavity. It provides direct evidence that the human is adapted for much more effective retronasal smell than the dog and other mammals.

The retronasal route to the smell organ begins with the food or drink that comes into the mouth. There the food is moved about by the tongue as it is chewed (masticated). It is said that only the human can roll its tongue, an ability that may impart special capabilities in manipulating

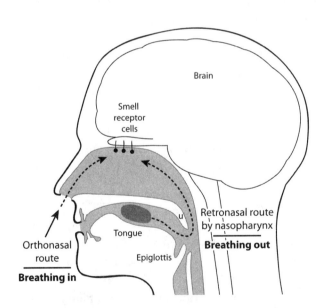

FIGURE 2.2 The head of a human
The arrows show the pathways in humans for sniffing smells by the orthonasal route and for sensing smells from the mouth by the retronasal route. Note that the pathways are relatively direct compared with those of dogs.

food as it is chewed and sensed. At the same time, the taste is sampled by the taste buds on the tongue and the back of the mouth and into the pharynx. When the chewer exhales, air is forced from the lungs up through the open epiglottis into the nasopharynx at the back of the mouth. There the air absorbs odors from the food that coats the walls and back of the tongue and that have volatilized from the warm, moist, masticated mass. Because the mouth is closed, the odor-laden air is pushed into the back of the nasal chamber and out through the nostrils, sending eddy currents within the nasal chamber up to the olfactory sensory neurons to stimulate them.

How might retronasal smell in the human compare with that in the dog? I know of no studies of how effective the passage of air is from the back of the dog's mouth through this long, narrow nasopharynx to the olfactory epithelium, so one can only speculate that the passage is less effective than in the human, where the distance is relatively short and the opening is relatively large. On this basis, one can hypothesize that the human nasopharynx is adapted for enhancing retronasal smell.

There are several reasons why retronasal smell may be especially important for humans, perhaps uniquely so:

1. With the adoption of bipedalism, ancestral humans became increasingly wide-ranging, migrating out of Africa to populate the world, concomitantly diversifying their dietary staples and their retronasal smells.

2. The advent of fire, at least 400,000 years ago, and possibly much earlier, made the human diet more odorous and tasty. From this time one can begin to speak of human cuisines of prepared foods, with all their diversity of smells. As early as the eighteenth century, James Boswell was speculating that what makes humans special is their invention of cooking. Richard Wrangham at Harvard University, in his work and recent book *Catching Fire: How Cooking Made Us Human*, has argued on modern evidence that prepared cuisines based on cooked foods are a defining characteristic of humans. Daniel Lieberman has provided much new insight into how this involved the mouth, pharynx, and retronasal route in his recent book *The Evolution of the Human Head* (for more on this, see chapter 17).

3. Added to the cooked cuisines were fermented foods and liquids, which made foods not only more diversified but also more intense in their tastes, smells, and flavors.

These developments occurred among the early hunter-gatherer human cultures and lasted through the last ice age. With the transition to agricultural and urban cultures around 10,000 years ago, human cuisines expanded and stabilized through the addition of products from domesticated animals, plant cultivation, use of spices, and complex procedures such as those for making cheeses and wines, all of which produced foodstuffs that especially stimulate the smell receptors in the nose through the retronasal route and contribute to complex flavors.

These considerations suggest the hypothesis that the retronasal route for smells has delivered a richer repertoire of flavors in humans than in subhuman primates, dogs, and other mammals. On this basis, I postulate that this system in the human brain played a much more important role in the evolution of early humans than has been realized, as well as a much more important role in our daily lives (including my daily home dinner).

CHAPTER THREE

How the Mouth Fools the Brain

I have claimed that smell is the main component of flavor, and that it is stimulated through the retronasal route. What is the proof?

Psychologists and food scientists answer this question by devising sophisticated methods to introduce food volatiles at the back of the mouth, synchronized with expiration, to show activation of smell responses by the retronasal route. But there is a simpler test, one that is often used by psychologists when visiting schools and giving demonstrations of scientific principles in science class. It's called the *nose-pinch test*, usually done with a piece of candy, but you can use any small, tasty morsel, even a dab of spicy sauce at the dinner table.

The Nose-pinch Test for Retronasal Smell

To perform this test, take the "tasty" bit and, while holding your breath, open your mouth and place the bit on your tongue while pinching your nose so that you can breathe only through your mouth. If you have good control, you can also prevent air from entering your nose by lifting the palate at the back of your mouth while you continue to breathe through your mouth. The requirement by any method is to prevent any air you are breathing out (exhaling) from going through your nose.

With the bit on your tongue, you will be able to identify certain senses. If it is candy you'll be able to detect the sweet taste due to the sugar. You will also know from the sense of touch that the bit is resting on your tongue, and where it is. If it is a morsel of food, your sense of touch will tell you whether it is hard or soft, or maybe hot or cool. If you sense any more than this, you are probably not holding your breath!

Now unpinch your nose, or lower your palate. Immediately you will experience the flavor of the candy or of the meat or other food or spice you may have chosen. The sudden whiff of flavor can be surprising. Almost without realizing it, you will have let the air shoot from the back of your mouth the short distance through the nasopharynx to the smell receptors in the nasal cavity to stimulate your sense of smell. The faintest whiff of outward breathing will produce this result. In fact, the diffusion of the volatile molecules in the air of the nasopharynx will account for much of the sensation.

Flavor Is Mostly Retronasal Smell

This experiment, and others by more scientific methods, tells you several key things about the role of smell in flavor:

1. It is clear that if there is no breathing out, there is no smell or flavor. You can confirm this by breathing in instead of out as soon as you unpinch your nose: there will be no sensation of flavor; it happens only with breathing out. This proves that the smell component of flavor is due exclusively to retronasal smell.

2. The ability to identify the type of flavor, such as lemon or strawberry, is clearly attributable to the sense of smell, which works in conjunction with the sweet taste and the senses of touch.

3. Perception of the retronasal smell is usually fused with the taste and touch so that we are unaware of it as smell. This is why the importance of smell in flavor has gone unrecognized for so long.

4. You have just proved that smell is the major component of flavor. Although sweet is a strong sensation associated with candies and other foods that contain sugar, sugar by itself is sweet and little else. The actual flavor of the candy depends on the sense of smell.

5. Even though by unpinching your nose you proved that the flavor of the candy is due to the smell you sense in your nose, when the smell is fused with the taste and touch of the candy in your mouth it appears to

come from the mouth. Not only is smell not recognized as a part of flavor, the flavor is not even recognized as coming from the nose. Rather, it is perceived as coming from the mouth. The mouth has taken all the credit!

The Mouth Takes All the Credit

Why should this be? Where else do we have a sense that is divided into two and one of them is hidden among other senses? Scientists are only beginning to realize that this is an interesting problem for psychology, for neuroscience in general—and for determining the food we buy and consume.

We do not know the answer, but it will surely involve the fact that when we ingest food, we put it in our mouths. Although we take this simple act for granted, when it is looked at from the perspective of animal behavior it is seen as being symbolically and practically a highly significant behavior. Something is taken from the outside world and inserted into the body. The animal has acquired it as a potential foodstuff—by killing another animal, or consuming some grass, or picking some fruit. The foodstuff looked as though it would be flavorful and nutritious, but it needs to be tested.

The touch system tells us the food is in our mouths; our attention is therefore focused on what is in our mouths as we make an assessment of it. That assessment involves, first, assuring ourselves that this is something we want to take into our bodies. Does it have anything unpleasant about it—a fish bone, a rotten taste? Is it too salty or sour or possess a bitterness that might signal a toxic or poisonous substance? Is it too hot or cold? The list is long. Second, if it passes the aversion test, is it something we like? If so, we lavish even more attention on it, because a desired flavor is one of the greatest of human joys.

Because our attention is on the food in our mouths, our perception follows our attention. It appears that this is what happens to retronasal smell. The attention is on the mouth, and the smell perception, due to the molecules released from the mouth, is merged with the other senses in making the assessment.

Because the smell in the nose is sensed as if it is coming from the mouth, it is a property of the nervous system known as *referred sensation*, which occurs when a sense appears to be in one place but actually arises in another. In this case, the sensation is not only referred to another place, but hidden in the senses from that place.

How to Assess Smell Within Flavor

Does it matter? Assessing smell within flavor gives us new insight into how complicated our sense of smell is. We have seen that smell is actually two senses, one for breathing in and one for breathing out. The one for breathing in is the sense of smell we all recognize as a single sense emanating from our nose. However, the one for breathing out is never recognized as a distinct sense, because it is always fused with two other senses, taste and touch, to form a third sense—flavor, which is referred to another part of the body, the mouth. Retronasal smell is unique among all our senses in these respects. If, as Brillat-Savarin maintained, "Tell me what you eat and I'll tell you what you are" is true, it follows that retronasal smell is the hidden factor that makes us what we are.

From the point of view of flavor, the unique properties of retronasal smell pose new challenges in understanding the flavors of the food we eat. Our nose-pinch test shows that you can explore these challenges yourself to reveal more clearly the contribution of retronasal smell to the flavors of what you are consuming. Separating retronasal smell from overall flavor would seem to be especially of interest to a food scientist or a gourmet cook, who wants to understand the flavor of the dish that is being cooked by analyzing each sensory property and how it arises from the molecular composition of the dish. This is one of the avowed aims of the new field of "molecular gastronomy." I think we can see the beginnings of a close collaboration with neurogastronomy.

Given that retronasal smell is stimulated by the release of smell molecules from our ingested food, one would suppose that we would sense the smell of our food separately from the stimulation of the taste buds on the tongue and the stimulation of touch receptors in the mouth. The nose-pinch test has shown that the smell can be sensed separately only by holding your breath and then releasing it to isolate the smell component of the flavor. Otherwise, the only thing you sense when eating your food is the flavor as a unified sense.

Flavor is thus the unified percept that we strive for in cooking a dish. We try to determine whether it was made this time with too much or too little salt, or enough curry, or was cooked too long or not enough, or any of the dozens of variables possible. Smell has the property of being, in general, "synthetic"; that is, a mixture of several smells makes a new unified smell. It is not "analytic," the way taste is: sweet and sour tastes sweet and sour rather than being a new unified taste.

The synthetic quality of smell challenges us to explain how the smell pathway discriminates its stimulus molecules, so that we can understand its contribution to flavor. I start with an introduction to the basic types of flavor molecules in the foods we eat that activate the sense of retronasal smell.

CHAPTER FOUR

The Molecules of Flavor

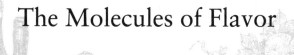

As Jean Anthelme Brillat-Savarin predicted,

> The number of tastes is infinite, since every soluble body has a special flavor which does not wholly resemble any other. . . . Up to the present time there is not a single circumstance in which a given taste has been analyzed with stern exactitude. . . . Men who will come after us will know much more than we of this subject, and it cannot be disputed that it is chemistry which will reveal the causes or the basic elements of taste.

There are indeed men and women who have come after, who definitely do know about the chemistry of taste in terms of the molecular composition of the foods we consume. It will be useful to review some of the main types of these molecules in order to relate them to the brain mechanisms that create their flavor.

As mentioned in the introduction, studying food odors and flavors is the job description of several kinds of scientists with distinct but overlapping expertise and aims. *Psychologists* and *psychophysicists* have a natural interest in testing human perception. *Organic chemists* who work in the flavor and fragrance industries routinely test the odors of the chemicals they are synthesizing in order to develop new flavors and fragrances. *Food scientists* try to understand how a particular food gives rise to a particular flavor, simulating or heightening it with artificial odor molecules. They classify odors on an empirical basis related to particular types of behaviors or to specific types of food flavors or perfumes. To do this, they use modern instrumentation to pull apart the molecular composition of a food to relate it to the flavor perceptions of that food.

To illustrate the complexity of the smell of food, coffee is a good example. It has been calculated that the taste and aroma of coffee are due to some 600 ingredients. How do we relate a particular ingredient to a particular aroma? The main instrument available at present is the gas chromatograph/mass spectrometer (GCMS), widely used to study the composition of different materials. A gas containing vaporized molecules of the substance under study is inserted through a long tube in the GCMS, so that the different molecules are deposited on a collecting surface in accordance with their molecular weight. The different molecules give off different wavelengths of light when illuminated. This enables them to be identified by the mass spectrometer, which can measure a substance down to a concentration of around one part in a million. The readout consists of peaks at different sites (allowing identification of the substance) and with different peaks (showing the amount of a given compound).

The GCMS is a valuable tool for analyzing the molecular composition of a substance. However, the odor world is too clever. There is often a poor correlation between the heights of the peaks in the MS readout and the strength of the odor from that compound. Often the main odor qualities of a foodstuff are due to molecules which are scarcely detectable in the MS.

A good analogy for why this should be can be found in gemstones. It is usually the case that a precious stone, such as a ruby or sapphire, owes its striking color not to its main mineral constituents but to a trace element in the stone. So it is with odors: a given foodstuff may owe its most characteristic odor to a trace amount of a type of odorous molecule. The experiments of the psychophysicists tell us that we can detect an odorous compound even though it is not detectable in the GCMS. Thus we humans, despite the bad publicity about our smell abilities, have a sense of smell that is better than the most powerful molecule-detecting devices that our brains have been able to devise.

To understand how our brains construct perceptions of smells, we need to start with an understanding of the molecules themselves. Peter Atkins of Oxford University is one of the best popularizers of the molecular nature of things. His book *Molecules* gives a good sense of the chemical structures that make up our material world, many of which are able to give rise to smell perceptions. We summarize here some of this information to give a sense of the principles of how these different odor molecules are related, and how they are able to carry the information about their sources in the perceptions to which they give rise. This is as fundamental for molecular gastronomy as it is for neurogastronomy.

Fruit Odors

Humans have descended from a line of primates that ate primarily fruit, so it will be appropriate to begin by considering fruit odors.

Ethylene (*ethene*) (C_2H_4) plays a central role in the ripening of fruit. It is a volatile molecule, meaning that it can be released as a gas into the air. Its release is the signal that the peak of ripening is occurring. This may reflect an action of the gas in activating the metabolic processes involved in ripening, including softening of the cellulose of the cell membranes. As Atkins relates, fruit is often shipped unripe and then exposed to ethylene gas to bring it to ripening when sold. (Judging from the lack of natural flavor in most of the produce at the supermarket, this process is not very effective in practice.)

Apart from this role in food, ethylene is also produced during refining (known as *cracking*) of petroleum. It can form long chains, producing polyethylene, which is a good insulator because of the inert nature of the bonds. As such, it is the prototype of many other types of plastics. Atkins notes many similar examples of how a given molecule can be used in totally different ways.

Food scientists are carrying out detailed analyses of the dozens or even hundreds of compounds that may be present in a given food. They want to know the way each compound contributes to the flavor as a volatile substance released from ingested food in the mouth to stimulate the nose by the retronasal route, and at different stages during ingestion, chewing, and swallowing.

An example of this kind of study is an analysis of the aroma of a banana carried out by D. Mayr and his colleagues at the Nestlé Research Center in Lausanne, Switzerland, in 2003. This is described in the article as a "breath by breath analysis of volatiles released in the mouth during eating of ripe and unripe banana using Proton Transfer Reaction-Mass Spectrometry." In other words, they wished to show the production of smell molecules in successive breaths during chewing of unripe as opposed to ripe bananas. The aim was to gain insight into what signals the consumer is getting to detect and eventually decide whether a fruit is ripe.

The retronasal smell of the ingested fruit morsel comes with the first outward breath, carrying the odor molecules from the morsel through the back of the mouth and up to the olfactory sensory cells in the nose. The smell consists of dozens of kinds of volatile molecules released from the banana cells. Chief among these are the simple alcohols, such as the two-carbon ethyl alcohol familiar in alcoholic drinks. (Some suggest that the craving of our primate ancestors for eating ripe bananas to excess disposed humans to alcoholic drinks and susceptibility to alcoholism.)

These simple alcohols give rise to a sensation that we call *sweet*, but this is only in analogy with the sweet sensation arising from the action of sugar in the taste pathway. We assume that the real meaning of the term *sweet* has to do with the taste of sugar and, by analogy, to the smell of the alcohol. But is this a result of our use of language, rather than an intrinsic property of either the sugar or the alcohol? Does a mute primate distinguish the two sensations? Did early prelanguage humans distinguish them? As language developed, how did the terms describing these two sensations emerge? These considerations indicate that relations between smell and language are a rich field, which we explore further in chapter 24.

Analyzing Flavor

The study of how flavor is produced from eating a food can be said to have begun with Brillat-Savarin's section "Analysis of the Sensation of Tasting":

> He who eats a peach . . . is first of all agreeably struck by the perfume which it exhales; he puts a piece of it into his mouth, and enjoys a sensation of tart freshness which invites him to continue; but it is not until the instant of swallowing, when the mouthful passes under his

nasal channel, that the full aroma is revealed to him. . . . Finally, it is not until it has been swallowed that the man, considering what he has just experienced, will say to himself, "Now there is something really delicious!"

In another section, "The Supremacy of Man," he analyzes the process even further, with special attention to the movements of the lips, mouth, tongue, and throat that appear to him to be unique to humans:

[M]an's apparatus of the sense of taste has been brought to a state of rare perfection; and, to convince ourselves thoroughly, let us watch it at work.

As soon as an edible body has been put into the mouth, it is seized upon, gases, moisture, and all, without possibility of retreat.

Lips stop whatever might try to escape; the teeth bite and break it; saliva drenches it: the tongue mashes and churns it; a breathlike sucking pushes it toward the gullet; the tongue lifts up to make it slide and slip; the sense of smell appreciates it as it passes the nasal channel, and it is pulled down into the stomach . . . without . . . a single atom or drop or particle having been missed by the powers of appreciation of the taste sense.

It is, then, because of this perfection that the real enjoyment of eating is a special prerogative of man.

This pleasure is even contagious, and we transmit it quickly enough to animals which we have tamed.

In our day, food scientists pursue the detailed study of the eating process with sophisticated instruments to analyze both sensory qualities and motor activity.

The banana is probably the most analyzed of all foods by these methods, which seems appropriate. It is an icon of primate preference for fruit, and more bananas are consumed annually worldwide than any other fruit. It contains more than 300 volatile compounds. A typical study focused on two of the main types of these compounds, esters and carbonyls, as registered by proton-transfer-reaction mass spectrometry (PTR-MS). An artificial mouth that performed artificial chews was used to test for the release of volatiles from unripe and ripe bananas. It found that unripe fruit released relatively few ester and carbonyl volatiles, even with the highest chewing rates. By contrast, ripe fruit released many volatiles, and these carried the main attractive qualities of a ripe fruit, such as being fruity, apple, candy, floral, caramel, and yes, definitely "banana-like."

As a general rule, "fruitiness" is due primarily to ester molecules— that is, molecules with a double-bond oxygen on an internal carbon— that are produced when, through enzymatic action in the ripening plant, an acid is combined with an alcohol. In the natural state, as chewing of the fruit proceeds, there are added retronasal smells from the products of the actions of the enzymes in the saliva. These fill out the sensory profile of the fruit and provide clues to slight variations in degree of ripeness and varieties of a given fruit that are stored in memory in the brain and that guide future feeding preferences.

The volatiles are described in terms of the main notes (fruity, candy [sweet], banana), and secondary notes (cheesy, green, apple, pineapple, floral, chocolate, caramel, mushroom). Is this telling us there is a smell "image" for ripe banana that overlaps to some extent with the smell "images" for these other food objects? These specific molecules are the "signature" of a ripe banana. As we shall see, the brain percept of these molecules is a pattern in the brain that monkeys, and humans, like or crave in a naturally ripened delicious fruit.

Why Plants Are Smell Virtuosos

According to Harold McGee, "plants in general are biochemical virtuosos," giving off multiple kinds of aromas. McGee is a studious person with an engaging personality who could well pass for a professor of English literature, which is what he started out to be until sidetracked by a passion for the chemistry of food. Now he is widely recognized, along with others such as Shirley Corriher, as a pioneer in understanding the science behind cooking and the food flavors it produces. We met at a workshop where I gave a talk on smell images and the brain flavor system and he was one of the attendees, learning still more about food and the brain.

Through his books McGee has given insights into the molecular basis of foods and their flavors that are of practical benefit for everyone from food novices to professional chefs. Here are some observations from his book *On Food and Cooking: The Science and Lore of the Kitchen* on the molecules in plants that give them their virtuoso smells and flavors.

• *Green.* This aroma is faint until the plant tissue is torn apart, or cut, as by a knife, or chewed. These actions damage the cell membranes, causing an oxidizing enzyme called *lipoxygenase* (*lipid* = *fat, oxygenase* = *to break down by combining with oxygen*) to act on the fatty molecules that make up the cell membrane to break them down into small, volatile fatty acids, which are further dismantled by other enzymes in the cell contents. The aroma of *green*, therefore, is not primarily an intrinsic smell—that is, one given off by the natural leaf itself in any great amounts—but rather one produced secondarily on preparing or eating the plant.

• *Terpene.* These are five-carbon molecules that can take many forms. They are common constituents of plants as well as of fruits, herbs, and spices. They are referred to as *ethereal* because of the etherlike "lightness" of sensation they produce. The distinctive smell of pine trees in the forest is due to terpenes in the tree resins. They are highly volatile and therefore act quickly when raw vegetables are cut or chewed, and they are quickly lost in cooking. They are also highly reactive with each other and with other molecules.

• *Phenols.* These are six-member carbon-ring molecules with a variety of side groups hanging off them. Different phenolic compounds are responsible for the main "notes" of different herbs and spices.

• *Sulfur.* Sulfur-containing molecules often are produced by a plant for defensive purposes. They have an aroma with an "edge" that gives a "pungent" quality to the smell, often because of direct stimulation of touch fibers in the nasal membranes.

Herbs and Spices

Among the plant virtuosos, the herbs and spices reign supreme when it comes to smell and flavor. Since recorded history began, they have been delivering intense aromas to human cuisines. One can hypothesize that they may have been doing this in prehistoric times as well, perhaps even enticing the early humans into migrating out of Africa, as the plants must have been flourishing during those times.

Herbs and spices play dominant roles in the defining flavor elements of many traditional cuisines. Whereas most plants contribute nourishment as well as flavor, herbs and spices are added to a cuisine primarily for their flavor contribution. Strangely, they are almost completely ignored in public policy discussions of how to construct healthy diets.

Each plant actually consists of many volatiles that contribute to its characteristic odor. As with other food sources, the characteristic aroma is due to a few key molecule types. Because of their strong stimulating abilities, these molecules are often used in studying the brain responses to odors.

Finally, as Atkins notes, *2-tert-butyl-4-methoxyphenol* (BHA) ($C_{11}H_{16}O_2$) is an antioxidant by virtue of the fact that it opposes the ability of oxygen to degrade molecules. These actions have been recognized only in recent years. Many spices, such as sage, cloves, rosemary, and thyme, are rich sources of compounds resembling BHA. Already in the Roman Empire, spices were used for their preservative and antibacterial properties as well as for flavoring.

Why Meat Tastes Like Meat

The great leap forward in the development of human cuisines, and the associated emergence of human culture and language, was the discovery of the use of controlled fire to cook foods, as described by Richard Wrangham in his book *Catching Fire: How Cooking Made Us Human*. The greatest effect was achieved by cooking meat.

There does seem to be something about cooked meat that humans instinctively like. It is mostly due to the smell, both the aroma breathed in through the nose (orthonasal smell) and retronasal smell from ingested food in the mouth associated with flavor. The most attractive volatile molecules from cooked meat are produced by the "Maillard reaction." The special flavors produced by cooking meat are described by McGee: "Raw meat is tasty rather than flavorful. It provides salts, savory amino acids, and a slight acidity to the tongue, but offers little in the way of aroma." This means that for the dog and other carnivores, not only does the long nasopharynx limit the aromas coming from the mouth, but the raw meat gives off little flavor to begin with.

When humans invented cooked meat, they not only added flavor to the meat, but they had the short nasopharynx to enjoy it. McGee continues:

> Cooking intensifies the taste of meat and creates its aroma. . . . [Up to] the boiling point . . . its . . . flavor is largely determined by the breakdown products of proteins and fats. However, roasted, broiled and fried meats develop a crust that is much more intensely flavored, because the meat surface dries out and gets hot enough to trigger the Maillard or browning reaction. [These aromas] have a generic "roasted" character, but some are grassy, floral, oniony or spicy, and earthy. Several hundred aromatic compounds have been found in roasted meat.

Thus, in contrast to raw meat, the properties produced during cooking are varied and dramatic, and at the heart of much of human flavor.

Heat causes the muscle cells of meat to break down, releasing the molecules that smell meaty and also breaking down other protein and fat molecules to form new molecules—particularly esters, ketones, and aldehydes—that give the meat new smells such as fruity, floral, nutty, and grassy. Fat is in fact responsible for most of the characteristic flavors of different meats. These smells are produced at relatively moderate cooking temperatures. One can imagine that these new smells were a delightful discovery by early humans, enough by themselves to motivate the development of cooking meat as a desirable part of the human diet.

At high temperatures that produce a crust on the meat, a level is reached at which new reactions take place, especially between sugar molecules from the breakdown of carbohydrates and amino acid molecules from the breakdown of proteins. The reactions that form these new molecules are called *Maillard reactions*, after Louis-Camille Maillard, the French scientist who first described them almost a hundred years ago. These reactions cause the dark brown color of the surface of seared meat. The molecules produced are highly volatile and have even more intense smells of the meaty, floral, and fruity varieties.

With increasing heat there comes a trade-off between the volatile smell-producing molecules that are characteristic of a given meat (or vegetable, for that matter) and the new molecules produced by the Maillard reaction. (According to McGee, the latter are more intense but less individualized.) Expert cooks explore these trade-offs with their different methods of lower heat preparation (such as boiling) and highest heat (frying).

The aroma of cooked meat can be sensed before cooking by inhalation through the nose; however, these are smells only from volatized molecules at the surface of the meat. In contrast, the full amplification and diversification of aromas from all the molecules within the meat produced by cooking can be sensed only when released from within the mouth, and when the other sensory pathways involved in flavor are activated in parallel, especially the pathways for taste and touch.

The texture of cooked meat is due to the balance between the proteins of its muscle cells, the collagen that makes up its connective tissue, and the water contained in or around both. This balance alters during heating. McGee describes three stages: rare, characterized by early juiciness when the muscle fibers begin to coagulate; medium-rare, when the collagen fibers become denatured and shrink, squeezing out water and making the meat drier and chewier; and well-done, when the collagen softens into gelatin while the muscle fibers become looser. The balance between these factors obviously depends on the type of meat, the amount of heat, and the duration of heating.

Dairy Smells

Dairy products, such as butter and cheese, arise from the milk of domesticated cows and goats, so it is assumed that they became a part of human cuisine only during historic times, in the past 10,000 years. The flavors of butter and cheese, largely because of their aromas, have made them an integral part of the diets of many human cultures. Here we follow Atkins in noting a couple of molecules responsible for that attraction.

Butanedione ($C_4H_6O_2$) has a cheesy, buttery, acrid smell. It is a ketone (also called *diacetyl*) because it contains a carbonyl group $C=O$; this type gives rise to a wide range of smells and flavors. Butanedione has a cheeselike smell, which is responsible for the main flavor of butter. It also contributes to sweat and armpit odor. Psychophysical testing has shown that these types of odors can give rise to either the cheeselike or armpitlike perception, depending on the context.

Butanedione is added to margarine to give it its buttery flavor. *Linoleic acid* ($C_{18}H_{32}O_2$) is the main fatty acid in many vegetable oils, such as cottonseed, corn, soybean, and rapeseed oil. It is also used in margarine, shortening, and salad and cooking oils. It has little smell by itself. It is hydrogenated (by bubbling hydrogen through the linoleic oil) to prevent the fat from being oxidized; that is why margarine does not get rancid. However, the hydrogenation leaves the margarine white, so carotene molecules are added to restore the yellow color. Butanedione is added to give it a butter odor. This is a good example of how an understanding of molecules can give insight into the world we live in.

I remember when growing up in Iowa during World War II the dairy farmers insisted that margarine did not taste like real butter, so it had to be sold in its white state in a plastic bag, with the color in a separate capsule, which we had to break to knead in the color. This nonsense ended when, at a conference of the dairy industry, the butter served at the closing banquet was later revealed to have been margarine—and no one had complained. This shows that the appearance of a food influences its flavor, as we shall see in chapter 15.

In addition to these generic smell molecules, cheese contains molecules reflecting the grass that the cows consume. These can be shown to have specific effects on the aromas and flavors of the cheese. A particularly careful study was carried out in 2004 by Stephania Carpino, Guilelmo Licitra, Terry Acree, and their colleagues in Sicily. For her thesis work, Carpino stationed herself in the pastures and charted the specific types of grass that the cows ate through the day and through the season. She correlated this with the GCMS peaks and showed that the molecular composition varied with the type of grass and its maturity in different seasons. She then carried out perceptual testing to show that subjects—and presumably consumers—could distinguish the different cheeses produced.

Licitra is head of a dairy cooperative in Sicily that is using modern dairy methods to a practical end: to produce traditional cheeses with enhanced flavor qualities. Similar attempts are being made in many countries—including, fortunately, more and more in the United States—to counter the bland produce available in mass-market supermarkets with foods that carry the full sensory qualities of locally produced items.

———————————————

In sum, each food has its characteristic molecular composition, modified by how it is prepared. By themselves, foods have no flavor. They are the raw materials out of which the brain creates flavor.

PART II

Making Pictures of Smells

CHAPTER FIVE

Smell Receptors for Smell Molecules

Most explanations of the brain systems involved in flavor start with the sense of taste. However, we have made the case that, paradoxically, smell is more important for taste than taste is, so we start with smell and give most of our attention to it. This is the first step toward a scientifically based neurogastronomy.

The sense of smell begins with the action of smell molecules on the receptor molecules in our nose. Here lies a second paradox. Smell molecules in our food, as described in the previous chapter, have been the subject of study for many years. Companies producing the ingredients of processed foods employ armies of organic chemists, who characterize the stimulating properties of thousands of different chemicals and try to relate them to the perceptions they arouse. But it takes two to tango: these smell molecules, and the receptor molecules with which they interact. Until 1991 nothing was known about the receptor molecules, so the critical molecular basis of the smell response and the subsequent perception was missing.

In 1991 the receptor molecules were discovered, making it possible to start the tango with both partners together. However, because of a host of difficult issues with the receptors, the scientific field was very slow to take up the problem of how the two partners interact. Subsequently, when molecular gastronomy came onto the scene in the late 1990s, information about the receptors was barely starting, and it is still slow in coming because of the many problems that still have to be solved. Thus the paradox: you might think that molecular gastronomy is molecular because it concerns both the smell molecules and the receptor molecules, but in fact so far it is mainly concerned with the perceptions of the smell molecules.

For neurogastronomy, the receptor molecules are a critical part of the smell and flavor system in the brain, so our interest begins with the interaction of the smell molecules and the receptor molecules in the nose. From this perspective, we are describing a subfield that we might call *molecular neurogastronomy*. As the new kid on the block, it lacks the richness of lore and knowledge about foods and their flavors that gastronomy and molecular gastronomy embrace. However, all this knowledge will merge increasingly to delight our culinary as well as scientific appetites.

Everything that follows in building up the sensations of smell and flavor starts with this critical and fascinating meeting between the molecules that are in our food and those that are in our receptor cells. How does it work?

The Concept of a Lock and Key

Imagine that a molecule is like the key to your house. Run your finger along the jagged edge that fits exactly the complementary edge inside the lock. When you insert the key into the lock the two come together, you can turn the key, release the lock, and open the door. This "lock-and-key" concept has been used by biologists for more than a century as the model for how two molecules interact with each other. When the key turns the lock, the molecule changes its internal structure. This change results in a microkick delivered to a neighboring molecule, which starts a cascade of microkicks to other molecules that results in the cell doing what it is supposed to do.

An odor molecule is made up of different types of atoms that give it an irregular structure. It functions as the key. What does the lock look like, and how does it work? This is one of the big questions in the modern science of smell. Solving this problem has given us a remarkable insight: how the information contained in the odor molecule is represented by an image in the brain.

48

Challenges

Studying how a smell molecule activates a smell receptor is therefore part of the larger enterprise of understanding how the information carried in a sensory stimulus is transformed into a representation in the nervous system. This process is best understood in vision, where individual photons activate rhodopsin molecules in the photoreceptor cells in our retinas. It is also well studied in hearing, where sound waves are converted first into vibrations within our inner ears, which activate the hair cells in our cochleas. In each case, the optimal stimulus is known and can be controlled very precisely.

Sensory stimulation of smell receptors is much more difficult. We do not have the advantage of being able to "see" or "hear" the stimuli we are delivering. Monitoring a smell molecule in the air has to be done indirectly by instrumentation. The receptor cells are hidden inside the nasal cavity and are hard to get at in order to record their responses. The cells are easily fatigued by repeated stimuli (just as we rapidly adapt to an odorous environment), so testing has to proceed slowly. We usually do not know beforehand which odors will activate a given cell; finding this out among the thousands of possible odors takes a lot of time. Even when we identify an odor, we have to figure out what other odors are within the receptive range of this particular cell. All this work is carried out with orthonasal smell, which can be controlled by puffs of smell, but the basic mechanisms will apply to retronasal smell as well.

Features of Smell Molecules

The keys, in our analogy, are the smell molecules, many of them described in the previous chapter. They range from small molecules to larger musks and pheromones and even fragments of proteins given off by the body that are borne on particles in the air. This panoply of smell molecules is particularly relevant to the smells we obtain by sniffing in. However, our problem is simplified because retronasal smell molecules are mainly the smaller molecules that volatilize (evaporate) from the liquids and foods released from within our mouths.

49

What part of these small molecules is the part that stimulates the receptors? If we know that answer, we can begin to understand how that information gets represented in the brain. It has been known by organic chemists and shown by a host of physiological studies that changes in a single molecular feature, a single part of the molecule such as a single atom, can change smell perception. The fundamental bits of knowledge contained in smell molecules and represented in the brain are therefore likely to be individual features of a smell molecule. These are of several types.

I will build on chapter 4 to describe them. These features are what link molecular gastronomy to neurogastronomy.

Most obviously, odor molecules may vary by their length; thus, straight-chain (called *aliphatic*) molecules may vary from having one to a dozen or more carbon atoms in their backbone. A second difference resides in a terminal functional group. Many odor molecules have a terminal group that defines whether it is an alkane, acid, or aldehyde, each of which is associated with characteristic smells. As we saw, these are found in many kinds of foods. With these first two features, we can define a homologous chemical series as one consisting of molecules with the same terminal functional group (hence *homologous*) but with chains consisting of different numbers of carbon atoms (hence *series*).

A third difference is whether a *functional* group is contained within the carbon atom chain, such as an oxygen atom in a ketone. A fourth difference is in a *side group* attached to the side of a carbon chain, such as a phenol ring. A fifth difference is in the *chirality* of the molecule— that is, whether it is right-handed or left-handed. A sixth difference is in the geometrical *shape* of the molecule, such as the ringlike conformation of a terpene. A final difference is in the overall *size*.

The ability to detect these differences of single atoms within smell molecules makes the olfactory system one of the most sensitive molecular detector systems in the entire body. These features are at a much finer level than in the immune system, for example, where an antibody interacts with an antigen site consisting of dozens of individual molecules.

The Race for the Smell Receptors

What kind of receptor molecule could detect such fine differences between molecules, and in addition, between thousands of different molecules? This was one of the great mysteries of science. The earliest ideas were by industrial chemists working with organic molecules. They posited that particular molecular features interact with unknown receptors. These ideas merged after a time with the larger field of study of what are called *structure-activity relations* in molecule-molecule interactions. These studies evolved into Quantitative Structure-Activity Relations (QSAR), a standard approach by pharmaceutical companies for developing new drugs.

These studies revealed a daunting complexity of odor molecules and their relation to perception. On the one hand, a series of molecules with similar molecular features—such as alcohols, esters, and aldehydes— might give a series of similar perceptions. On the other hand, similar molecules could give quite distinct perceptions. This was baffling for the classical QSAR approach.

On the basis of an analysis of molecular shapes, John Amoore, an Oxford biochemist, suggested that the receptors are tuned to the shapes of the odor molecules; this is the so-called *stereochemical theory* of odor receptors. But the nature of the receptors, the specific kinds of proteins in the cell membrane, remained unknown.

The first breakthrough came with biochemical research by a former student of mine, Doron Lancet, in Israel. In 1985 he galvanized the field by showing that odors stimulate an enzyme called *adenylate cyclase*, which produces a well-known cell signaling molecule called *cyclic AMP*. It was already known that cyclic AMP occurs in a signaling pathway that starts with a receptor that gives a microkick to a so-called G-protein, which forms a large class of G-protein coupled receptors (GPCRs). Lancet drew on his earlier training as an immunologist to predict that a large number of different receptors—as many as 100 to 10,000—would be needed to encode all the different types of odor molecules, a prediction that turned out to be surprisingly close. He also drew on predictions by another former student, John Kauer, and earlier work by Kjeld Døving in Norway, that there would be a "combinatorial relation" between receptors and odor molecules; that is, a given receptor could interact with many odor molecules, and a given odor molecule could interact with many receptors.

Suddenly a number of laboratories in molecular biology realized that discovering the identity of these receptors was one of the hottest topics in biology. The race was on. One by one, the intermediate "microkicking" molecules in the odor-signaling pathway in the receptor cells were cloned and sequenced and identified by their pharmacological properties: the G-protein, the adenylate cyclase, and a protein that forms a channel activated by cyclic AMP that lets in charged particles to give the electrical response. The first recordings of odor responses of isolated receptor cells were made by Stuart Firestein, then a graduate student in Berkeley. The electrical responses in the cilia themselves were recorded. It was an exciting time. The whole sensory cascade was coming into view.

The whole cascade, that is, except for the receptors. Repeated attempts to identify the receptors met with no success. It became obvious that this was probably going to come from a major molecular biology laboratory with the needed expertise and resources. And that is what happened.

A Beautiful Experiment

Linda Buck was a postdoctoral fellow in the laboratory of a leading molecular biologist, Richard Axel, at Columbia University. She had been there for several years, working on several projects on endocrine receptors and on antibodies in the immune system. She began to read about smell and became fascinated by the field and by the problem of finding the receptors.

A few years previously, a technique called the *polymerase chain reaction* (PCR) had been invented, which made it possible to expose any given tissue in the body to a tiny part of a gene suspected to be present. The tiny part acted as a "probe" to recognize that whole gene and any similar genes if they were there, and to amplify them by repeated probing so that you had enough to recognize them and do something with them. You use this basic "probe" technique to fish out a text from the world literature. For example, if you enter "We hold these truths" into Google, you will recover the entire Declaration of Independence (you can try doing this with other phrases: it is remarkably effective). So PCR was just the right tool for going after the olfactory genes in the tissue in the nose, but nobody had yet made it work for that purpose.

On the basis of the prediction that the olfactory receptors would be a large subfamily of the G-protein coupled receptors, Buck designed her probes to fish out any possible new members of that family in the olfactory epithelium of a rat. She used snipping enzymes to cut her collected sequences in such a way that it produced a set of stripes on a biochemical gel showing that indeed there was a new and very large family of genes belonging to the GPCR family localized in the olfactory epithelium.

As she tells the story, at first she couldn't believe the results and put them away overnight. However, the next morning she realized that they were real, and she showed them to Axel, who was also elated. Reporting in 1991, they dubbed them the long-sought *olfactory receptor genes* that carry the genetic code the cell uses to make olfactory receptor proteins. They deduced that the family could contain as many as 1,000 members, in the range that had been predicted.

Discovering the olfactory receptors not only solved the problem of the smell receptors, but opened the door to a new era of research in the olfactory pathway. For these advances, the Nobel Committee in 2004 recognized Buck and Axel for their discovery, one of the greatest in the history of biology.

The Concept of an Odor-binding Pocket

How do these receptors fit with the lock-and-key model? Several years earlier, we and others had started to think about this problem. This led to the hypothesis that the interaction takes place not in a narrow lock that responds to only one key, but in a larger space called a *binding pocket*, analogous to that of many other receptors for signaling molecules between cells. We further hypothesized that a given receptor cell might carry just one type of receptor, which would require that the receptor does not have a narrow affinity for a given molecule, as with the neurotransmitter receptors, but rather a broad spectrum of affinities reflecting the known broad odor responses of olfactory receptor cells. The nice thing about this hypothesis is that it makes the olfactory receptor cells similar to photoreceptor cells in the retina, each of which carries only one of the three types of color receptors, but with a broad responsiveness to different wavelengths of light.

The "one cell–one receptor" concept was experimentally established by work in Richard Axel's laboratory and has received direct support from single-cell studies in the rat by Linda Buck's laboratory, then at Harvard University, and in the insect by studies by John Carlson at Yale University.

The problem is thus to account for a binding pocket that is analogous to that of other transmitter receptors and photoreceptors but can interact differently with different odor molecules. Thus, in contrast with the traditional lock-and-key metaphor for enzymatic and receptor specificity, the olfactory receptor is hypothesized to function by a broad affinity mechanism. This is an example of a new concept of receptor activation involving broad receptor-signal molecule interactions.

Interactions of Smell Molecules and Smell Receptors

The standard way to study receptor-signal molecule interactions in the drug industry is to use genetic tools to "express" the receptor (the technical jargon for *make it appear*) in a carrier cell, and then test it against different signal molecules and different potential drugs that block or enhance it. By genetic engineering, each amino acid in the chain that makes up the receptor can be changed at will, enabling the scientist to determine which amino acids are essential to the function of the receptor and which interact specifically with different sites (determinants) on the signal molecule.

Unfortunately, olfactory receptors have been very difficult to express experimentally in carrier cells. The breakthrough came in 1998 from my former student Stuart Firestein, his graduate student Haixing Zhao, and their colleagues at Yale and Columbia. They played a molecular trick by attaching a particular receptor gene to a virus, which they then poured over the sheet of receptor cells in the nose of a rat. This enabled the virus to infect all the cells so that they would all "express" that particular receptor (known as the *I7 receptor*). They could then record from any cell and find out which odor molecules could excite it. Among some 200 odor molecules they tried, the aliphatic (straight-chain) aldehyde composed of a chain of eight carbon atoms (C_8, octanal) was preferred, with decreasing responsiveness to flanking members in the series with chain lengths from C_6 to C_{10}.

This implied a binding pocket selective for at least two features: terminal functional group and chain length. To test this hypothesis, Michael Singer, a Yale undergraduate and then graduate student in our laboratory, carried out a molecular modeling study. A computational modeling analysis in 2000 of the I7 model receptor showed that, with blind automated docking, octanal interacts within a binding pocket as hypothesized. These studies thus identify four types of stimulating features on odor molecules: functional group, chain length, molecule size, and molecule shape.

A critical test of the computational results is the extent to which they reproduce the binding preferences of I7. There was close agreement: the model showed the same relative preference for octanal over the flanking members of the aldehyde series that the experiments showed. A parallel study from Firestein's laboratory demonstrated how the odor molecules that interact with I7 are tightly constrained at their functional head but can fit more loosely at their tail ends. It is presumed that this accounts for the graded affinities of these different molecules for the I7 receptor.

The combined result of these experimental and computational studies thus supports the hypothesis that odor determinants interact within a binding pocket in the olfactory receptors that is similar to that of other GPCRs, but with a set of critical sites that varies with different receptors and can show graded interactions with the determinants of different odor molecules.

Combinatorial Interactions Between Smell Molecules and Smell Receptors

Independent experimental testing of the hypothesis was provided in 1999 by the experiments of Bettina Malnic and colleagues in Linda Buck's laboratory. They first recorded physiologically the odor spectra (the range of different odor molecules that give responses) of isolated olfactory sensory neurons. They then sucked out the contents of the nucleus, which contains the genes, with a fine pipette and recovered the gene of the receptor by the PCR method. They could thus show the differing strengths of interactions between known receptors and different odor molecules. These and other experimental and theoretical studies support the hypothesis that the critical sites on the odor molecules (what we are calling *determinants*) are *transduced* (transformed) within a receptor binding pocket. The multiple combinations of amino acid residues and the multiple combinations of odor determinants within the binding pocket provide a fertile substrate for the combinatorial interactions that have long been posited to take place in encoding odor molecules, as described earlier.

These combinations are made even more subtle and complex by interactions at the receptors. The analogy here is with the drug industry. Pharmaceutical companies invest millions of dollars trying to find molecules that act as drugs to antagonize the response of a receptor to a neurotransmitter or to enhance it. We predicted that similar interactions could occur at olfactory receptors, except between different odor molecules competing for the same binding pocket. These indeed have been found, in which a response is seen to molecule A alone but not to molecule B alone; however, when they are presented together, the response to A is reduced because of the "silent" antagonistic effect of B on A. These interactions at the receptor binding pocket indicate that the complexity of a smell perception begins to be shaped at the first step in the smell response.

Understanding these interactions is obviously necessary for relating the molecular properties of the smell molecules to the perception of smell and flavor.

The Broad Molecular Receptive Ranges of Smell Receptors

Both experimental and computational studies reveal that the receptors have relatively broad response spectra that overlap with one another. The broad spectra have been termed *molecular receptive ranges* (MRRs) by a former student, Kensaku Mori, and Yoshihoro Yoshihara in Japan. This is in analogy with the spatial receptive fields (RFs) that are seen in cells responding to a visual scene in the retina.

The broad overlapping MRRs have been confusing to some, imply-ing a lack of specificity in the responses. The way that specificity can be achieved despite broad overlapping responses is easily explained by analogy with the color system in vision. The cone receptors in the retina contain three types of receptors. They are called *red*, *green*, and *blue*, terms reflecting the peaks of their sensitivities to different wavelengths of light. However, on either side of the peaks, all the receptors have decreas-ing sensitivities that overlap with one another across most of the wave-length spectrum. This means that at any wavelength there is a unique combination of responsiveness to the three receptors. This unique combi-nation is what gives rise to the unique perception we call *color*. A given combination stays the same regardless of the intensity of illumination, which enables us to discriminate color despite changes in light intensity.

We and others have suggested that a similar principle applies to smell, except that here one is dealing with the overlapping responses of hun-dreds of receptors, not just three. Nonetheless, it is the overlaps that enable us to identify unique combinations of receptor responses giving rise to a specific odor perception, no matter how weak or intense it may be. In this way, the brain achieves specificity of an ensemble of neuron responses despite lack of specificity of individual neuron responses.

The Multidimensional Nature of the Odor World

In contrast to color perception, in which wavelength varies along one dimension, odor molecules differ in all the ways we have indicated. Thus we say that "odor space" is *multidimensional*. This is because of the multiple combinations of the odor features that are possible; that is, a given molecule may belong to more than one dimension, such as a given functional group, a given carbon length, a given degree of saturation, and a given three-dimensional shape. Identifying molecular features is thus a necessary step in being able to characterize more precisely the odor world. As I will explain, this multidimensional nature of odor space poses special problems for its neural representation in the brain.

A Database for Olfactory Receptors

The mammalian genome is believed to contain some 30,000 genes. Three percent of them (approximately 1,000) are olfactory receptors, constituting the largest family in the genome. Just keeping track of all these receptors, comparing them and classifying them has presented a large challenge. For this purpose, my laboratory has created the Olfactory Receptor Database (ORDB). There are now more than 14,000 receptor genes and proteins in the database, representing the human, mouse, rat, dog, chimpanzee, invertebrate, fruit fly (*Drosophila melanogaster*), mosquito, bee, and roundworm (*Cenorhabditis elegans*) categories. (ORDB is available on the Web for inspection and download at the SenseLab Web site.)

CHAPTER SIX

Forming a Sensory Image

How does the brain represent our sensory world in order to serve as the basis for perception? This is one of the oldest questions in philosophy and a central question for modern psychology and neuroscience. It is also central for neurogastronomy.

Modern neuroscience has shown that other sensory systems construct spatial representations of their stimuli. The body surface is represented by a body map, also called a *homunculus*. The visual world is represented by visual images. In audition, different frequencies are represented in spatial "tonotopic" maps. In all these cases, the spatial representations start with the sheet of sensory receptor cells and are maintained up to the cerebral cortex, where perception occurs.

Research over the past half-century has shown that the brain also represents smell molecules by spatial patterns. These patterns are not present in the receptor cell responses we have just considered, but are formed in the first relay station, the olfactory bulb. This is the core concept in our theory of smell, because these patterns function as what we call *smell images* for further processing by the brain as the basis for smell perception. It may seem strange that a nonspatial stimulus received by the receptors in the nose, as we have just seen, can be represented as a spatial pattern specific for that odor molecule. The way this comes about has been one of the best-kept secrets of the brain.

What is the advantage of forming a neural image of the information carried in smell molecules? The idea of a smell image seems at first so strange that it is difficult to answer this question. It will be an advantage to take a brief digression and learn what is known about the formation of a "visual image," the most obvious image that we form of our sensory world.

Legions of neuroscientists and psychologists have studied how this happens in the brain. Our hypothesis is that by studying the neural mechanisms for setting up and processing a visual scene in the brain, we will learn principles that will apply to the neural mechanisms for setting up and processing a smell image in the brain. If this hypothesis holds water, we will gain a new perspective on how smells are represented in our brains and how these representations contribute to the perception of flavor.

You might assume that the best insights into how humans form visual images would come from studying humans themselves. However, as in many fields of biology, the best strategy when starting on a problem such as this is to find an animal in which the particular system exists in a simple form, so that experiments can be carried out that would not be possible in humans.

Discovering Lateral Inhibition

Such a system is found in the eye of the horseshoe crab (*Limulus polyphemus*). If you live near the ocean or vacation on a beach or visit an aquarium, you may see one of these creatures, like a low inverted bowl in the sand or in the shallow water. Look closely and you can just make out two small eyes in its hard shell. These simple eyes can detect only light or dark or shadowy moving objects, but that is enough to tell the crab the general time of day and warn it of predators in its neighborhood.

This creature does not seem to promise much insight into human vision, but it did to a biologist named H. Keffer Hartline, who worked at the Rockefeller Institute in New York City in the mid-twentieth century. Hartline really believed in the simple system approach. As quoted in "Horseshoe Crabs and Vision," he used to tell his students who wanted to study the neural mechanisms of vision, "[A]void vertebrates because they are too complicated, avoid color vision because it is much too complicated, and avoid the combination because it is impossible." Thanks to Hartline and many subsequent scientists, we have learned general principles about vision that are surprisingly relevant to the principles underlying other sensory systems, including smell.

Without going into the details of the receptor cells, we can summarize by saying that each one is contained in a microcartridge that takes in a small part of the visual scene. The combined array should therefore give a faithful representation of the pattern of light falling on the eye. Hartline tested this by moving a bright spot of light halfway over the eye, then abruptly changing it to a lower intensity in the other half. The responses were accordingly strong and weak. He then repeated the experiment, but instead stimulating simultaneously the two halves of the retina. Now he observed a dramatic effect: at the light–dark border, the cells responded more than expected on the light side and less than expected on the dark side. The neural image of this simple pattern did not represent the real pattern at all.

In summary, how strongly a cell fires depends on how active its neighbors are: a strong cell gets stronger, and a weak cell gets weaker.

Contrast Enhancement in Space and Time

This effect had actually been described in humans by a German physicist named Ernst Mach in the nineteenth century. He had noticed that when we view a light–dark border, such as a sharp boundary between two walls with different illumination, the contrast is enhanced by a lighter band on the light side and a darker band on the dark side. These came to be called *Mach bands*. You can see them yourself if you look for them. (The bibliography provides a site for you to look them up on the Internet.)

Hartline showed that Mach bands are present even in the primitive eye of *Limulus*. He further showed the mechanism that produces them: lateral inhibitory connections between the receptor cells. Through these connections, the strongly excited cells at the border more strongly inhibit the weakly stimulated cells, and the weakly stimulated cells more weakly inhibit the strongly excited cells. The mechanism is called *lateral inhibition*. The effect is called *contrast enhancement*, because the difference between the light and dark areas is enhanced at their boundary. In a general sense, contrast enhancement also is a kind of *feature extraction*, the enhanced response to specific spatial features in a visual scene.

This is contrast enhancement in space. Hartline's laboratory also showed that there is contrast enhancement in time. When there is an abrupt step increase in illumination, a single cell responds with a large increase in impulse firing, which rapidly declines to a steady level somewhat higher than before. The overshoot in impulse frequency is called the *phasic* response, in contrast to the steady *tonic* response. It shows that the nervous system is sensitive primarily to a change in the environment rather than to an unchanging steady input. This contrast enhancement in time is the counterpart to contrast enhancement in space. After the initial increase in stimulation, lateral as well as self-inhibition comes on to counterbalance the higher level of steady stimulation.

Forming Images in the Brain

In addition to contrast enhancement and feature extraction, lateral inhibition has other functions. The visual image is blurred from the dispersion of light as it passes through a lens; lateral inhibition reduces this dispersion, a process called *image reconstruction*. Another function is *gain control*—that is, the adjustment of how much amplification there is for increases or decreases in sensory stimulation. For example, *Limulus* responds over a wide range of light intensities. To cover this wide range and still have mechanisms that enhance sensitivity for very weak stimuli, there must be a reduction in sensitivity as stimulation gets stronger. This process is called *gain compression*. Gain compression is built into electronic amplifiers through feedback suppression in a manner analogous to feedback and lateral inhibition in *Limulus*.

So the neural image formed by the eye is not a faithful image of the actual scene the way a camera image obtained at low contrast is. It is instead an abstracted image, a high-contrast image, in which edges in the visual scene are abstracted and enhanced and the nonchanging rest of the field is suppressed. Similarly, the stimulation is highest when a new scene appears, and it settles down if there are no changes. This is why a horseshoe crab is sensitive to a shadow or an edge of light or dark that moves, so that it can be alert to predators or prey.

Lateral inhibition has come to be recognized as one of the most important principles in the organization of sensory systems. The findings in *Limulus* were confirmed in the mammalian eye by a remarkable scientist, Stephen Kuffler. He was a refugee from Austria early in World War II, escaping by skiing over the Alps. Later, at Johns Hopkins University, inspired by Hartline's results in *Limulus*, he set up similar experiments in the cat retina, recording from single ganglion cells—the cells that carry the output of the retina to the brain—while stimulating the retina in different places with a spot of light. Stimulation at the site of recording (in the center of the receptive field) would often excite the cell, whereas stimulation further away (in the "surround") would inhibit it, or the other way around: central inhibition and surround excitation. These experiments showed that the principle of contrast enhancement applies to mammalian vision, in a form that came to be called *center-surround antagonism.*

Image Processing by Animals and Machines

When you get an eye examination and are asked to see letters of decreasing size, how well you do depends on the lens of your eye, which can be corrected with eyeglasses. It also depends on lateral inhibition, which heightens the contrast between the dark letters and the white background. This kind of visual scene, however, is highly artificial. Animals in the wild normally experience visual scenes full of objects at low contrast; survival depends on identifying prey or predator within those scenes. How the retina does it will give us clues to how the nose performs its task of identifying odors within low-contrast odor backgrounds, whether they are prey or predator odors in the environment or volatile food odors in the mouth.

Peter Sterling and Jonathan Demb, visual scientists at the University of Pennsylvania, explain how this works in *The Synaptic Organization of the Brain*. Creating the optical image of a high-contrast scene requires only light that is of relatively low intensity; this is why we can read books even by candlelight. However, for the retina to create an image of a low-intensity scene requires lots of light. An example would be the natural scene of a sheep in a forest as might be viewed by a predator at dusk. A photometer scan across a photograph of this scene shows the small fluctuations in intensity that occur on a relatively even background. Identifying these fluctuations in dim light is difficult. Why?

To answer this question, Sterling and Demb quote Albert Rose, an expert in video engineering, who compares the retina to a black canvas on which the individual quanta of light (photons) paint a kind of pointillist picture—that is, photons hitting single pixels of the picture. To make a high-contrast picture by turning on or off single pixels requires only approximately one photon per pixel, so it can be done in dim light, but making a low-contrast picture with many shades of gray requires gradations among many photons hitting many pixels. Thus, seeing clearly a low-contrast picture requires a lot of light, as we all have experienced.

Lateral inhibition in the retina, by the means we have discussed, is fundamental to increasing the contrast. It is also needed to combat noise. The more shades of gray exist, the more photons are needed and the noisier the picture. The problem then is to increase the signal-to-noise ratio, a fundamental function in all sensory systems (including smell). One way to increase the ratio is to have synchrony in the center of the receptive field that smoothes out irregularities; this is done in vision by having electrical connections between the photoreceptors that pool many receptor responses together. In smell there are other mechanisms, as we shall see. The other way is to use the lateral inhibitory mechanism to reduce redundancy. Much of a low-contrast natural scene has a relatively constant level of illumination. Much of the information in it is therefore redundant and can be removed, leaving the local sites of change more enhanced.

In the field of image processing, this is called *predictive coding*, because values near the center are predicted to be higher than the mean of the surround. As the illumination grows dimmer, the surround maintains its predictive value by broadening, as shown by experiment and also by theory.

Vision thus provides us with the most thorough analyses of the properties of image building and image processing in a sensory system. Although the details of the systems may vary, the underlying principles are quite general. We will see that they give new insights into how smell images are constructed and processed to provide the basis for smell perception.

Feature Extraction in Our Sensory World

In summary, the eye sets up a two-dimensional representation of the visual world—a visual image. The advantage of this representation is that the nervous system can form circuits that process the image. The processing involves a fundamental role of lateral inhibition and contrast enhancement that transforms the neural image into a form that is most appropriate to the operations of the brain in building visual perceptions and beyond. Mach himself explained this best:

> [S]ince every (retinal) point perceives itself, so to speak, as above or below the average of its neighbours, there results a characteristic type of perception. Whatever is near the mean of the surround becomes effaced. Whatever is above or below is brought into disproportionate prominence. One could say that the retina schematises and caricatures. The teleological significance of this process is clear in itself. It is an analogue of the abstraction and the formation of concepts.

All the principles I have discussed—the initial image representation in a two-dimensional sheet, lateral inhibition, contrast enhancement, temporal transients, center-surround inhibition, and feature extraction—play essential roles in the formation of neural images in all sensory systems. In *hearing*, each nerve fiber from our ear that carries information has a "best" sound frequency. Lateral inhibition between fiber pathways helps sharpen that frequency response. In *touch*, our ability to discriminate between two closely spaced points (known as *two-point discrimination*) is better in our finger tips than on our abdomens. This is because of the higher density of innervation of the fingertip skin and the presence of lateral inhibition in the central pathways, which improves the discrimination. Similarly, by manipulating food in our mouths with our tongue, we carry out "feature extraction" on the *mouth-sense* of the food—whether it is smooth or rough, dry or moist, hard or soft, and so on.

And smell? We are now in a position to consider how all these principles apply to the representation of smells in our brains.

CHAPTER SEVEN

Images of Smell
An "Aha" Moment

From the olfactory receptor cells, the pathway to smell perception in the brain passes through a series of regions: the olfactory bulb, the olfactory cortex, and the orbitofrontal olfactory cortex. The first processing steps take place in the olfactory bulb. Because of its critical role in forming the smell image that is the major component of flavor, we will take several chapters to explain how it works.

The Olfactory Bulb

As its name implies, the olfactory bulb is shaped like an incandescent light bulb, sticking out in front of the frontal lobe of the brain. In comparison with the visual pathway, which starts in the retina and progresses through the thalamus to the visual cortex, it is as if the olfactory equivalents of all these structures are compressed into just the olfactory bulb. A big challenge is understanding all that goes on inside, and this requires getting familiar with the cells.

Figure 7.1 shows the olfactory pathway in a representative mammal such as the rat. When smell molecules bounce in and out of a receptor binding pocket in an olfactory receptor neuron (*orn*), all an individual cell "knows" is how much the features of the smell molecules have tickled its binding sites. Again, the greater the tickle, the more the cell responds by generating impulses. The code that it sends on to the olfactory bulb is therefore in the form of frequency of impulses in the olfactory nerves (*on*),

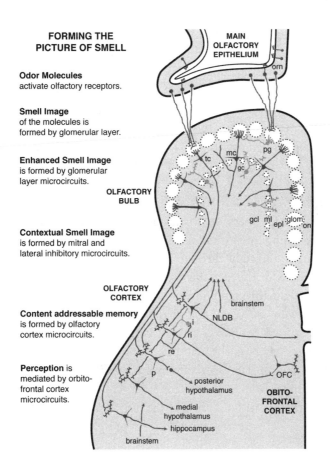

FORMING THE PICTURE OF SMELL

Odor Molecules
activate olfactory receptors.

Smell Image
of the molecules is
formed by glomerular layer.

Enhanced Smell Image
is formed by glomerular
layer microcircuits.

Contextual Smell Image
is formed by mitral and
lateral inhibitory microcircuits.

Content addressable memory
is formed by olfactory
cortex microcircuits.

Perception is
mediated by orbito-
frontal cortex
microcircuits.

FIGURE 7.1 The smell pathway

On the left, the series of operations to process the smell input from reception in the nose to perception in the cerebral cortex. *On the right,* the successive stages of the smell pathway that carry out these operations. In the **olfactory epithelium,** the main type of cell is the olfactory receptor neuron (orn).

In the **olfactory bulb,** the main types of cells are the mitral cell (mc); tufted cell (tc); periglomerular cell (pg); and granule cell (gc). The cells lie in different layers: olfactory nerve (on); glomeruli (glom); external plexiform layer (epl); mitral cell body layer (ml); and granule cell layer (gcl). In the **olfactory cortex,** pyramidal cells receive input from the olfactory bulb and connect to interneurons. Central fibers extending out to modulate the olfactory bulb cells arise in the nucleus of the horizontal limb of the diagonal band (NLDB). The **orbitofrontal cortex** (OFC) is represented by a single pyramidal cell to conserve space.

which does not tell much about which smell it is. This means that the code for smell molecules, the code the brain reads, must lie in the differences between the responses in the different cells.

In the olfactory bulb, the fibers from the several thousand receptor cells, all containing the same type of olfactory receptor, converge on a single site, called a *glomerulus* (*glom*). Depending on the species of animal, there are up to a thousand or so of these sites, each receiving its unique input. Connecting to each module within the olfactory bulb are some large cells called *mitral cells* (*mc*), named so because early histologists thought their cell bodies looked like a bishop's cap, or mitre. The mitral cells send their fibers on to the olfactory cortex. Together with smaller and more numerous versions of the mitral cells called *tufted cells* (*tc*), they provide the straight-through pathway. There are also numerous interneurons, cells with short branches, that are involved with local processing of the straight-through pathways. At the glomerular level they are called *periglomerular cells* (*pg*), and at the level of mitral and tufted cell output they are called *granule cells* (*gc*). Not shown, for simplicity, are parallel pathways that pass through the accessory olfactory bulb, often associated with pheromone reception, and a modified glomerular complex for special odor cues.

Through the pattern of its input and the interactions between its neurons, the olfactory bulb creates the code for representing the stimulating molecules.

How the Olfactory Bulb Represents Smells

The story starts at Cambridge University with Edgar Adrian, one of the great physiologists of the nervous system. After being a leader in the pioneering advances of the 1930s in the physiology of other sensory systems, he launched into a study of the olfactory system, the last major work of his career. He selected the brain of the hedgehog for his first study in 1943, showing electrical responses to natural odor stimuli. This was an era when biologists often chose species on the basis of behavioral considerations, whereas today there is an increasing focus on a few species available for genetic engineering. No one could doubt that the hedgehog burrowing through the ground lives primarily by its sense of smell!

One of Adrian's recordings became famous because its caption read: "Odor of decayed earthworm." The story goes that, in looking for a natural stimulus with which to test the hedgehog, Adrian found the shriveled remains of an earthworm in a dark and dank corner of his basement laboratory. By the time of these studies Adrian was a famous scientist with many administrative duties, yet he did all his experiments himself (and saved money by using natural stimuli).

Adrian next recorded from the olfactory bulb of the anesthetized rabbit. He put his electrodes in various parts of the olfactory bulb while stimulating with different odors. He found that the cells in different parts responded differently to different odors. We can do no better than quote from his paper of 1953:

So far then it looks as though Acetone molecules will produce an excitation coming mainly from the front part of the organ and from the particular groups of receptors in that area which have this specific sensitivity to it. A strong concentration may bring in other groups but, owing to the structure of the organ, there will always be critical regions where the concentration is only just strong enough to excite and here the specific effect will show itself. And there will also be critical times. At each inspiration the amount of material which enters the nose will increase progressively to a maximum and at the beginning and end of each inspiration the concentration is near the threshold value. The physical and chemical properties of the substance will therefore determine the time course of excitation. For instance, a large spike unit may have a specific sensitivity to Xylol. As the concentration of Xylol in the air is increased, other units will begin to come in during the later part of the discharge. With pyridine and eucalyptus the smaller spikes appear first and the large ones come later on. The result is that the photographic reproduction of the discharge has a characteristic shape for each substance, and this shape is reproduced with remarkable constancy each time the substance is presented to the nose.

The result is that the electrophysiologist, looking at a series of these records, could identify the particular smell that caused each one. We must not conclude that the brain identifies the smell by the same criteria but we can at least see how a great variety of smells might be distinguished without the need for very great variations in the receptors.

The basic concept that odors are encoded by spatial patterns and by their timing dates to this work of Adrian. Remarkably, he suggests how this differential encoding can come about "without need for very great variations in the receptors." As we have seen, very small differences in the molecular structure of the receptors accounts for their ability to respond differentially to specific features of the odor molecules. But Adrian knew the limitations of the knowledge he had at the time. His speculations were limited to the different shapes of multiunit recordings from assorted regions rather than being about actual mechanisms; he did not actually specify that these were "spatial patterns" in the olfactory bulb. He even cautioned against concluding that the brain recognizes smells in the way he could from looking at the different recordings from the different regions.

During my graduate work on the physiology of the olfactory bulb, I made the pilgrimage from Oxford to Cambridge to visit Adrian and discuss my experiments on the cells in the olfactory bulb and how they might relate to his studies. My adviser, Charles Phillips, cautioned me that Adrian was well known to be shy of visitors. Adrian courteously listened to my eager account of my work, but after a while he explained he was urgently needed elsewhere and began to make motions as if, literally, to run away. Our interview ended on the staircase of the old physiology building on Downing Street. My last question was what he thought was the most important problem to be solved in understanding the neural basis of olfaction. Over his shoulder came the reply: "Look to the glomeruli." With that, he disappeared. They were prophetic words.

A New Method for Mapping Brain Activity

Following Adrian's studies, electrophysiological studies made little progress in characterizing further the responses produced by different smells in the olfactory bulb. The problem was that, in contrast to the situation in vision and somatosensation (touch), where the experimenter knows where to put the electrode in a particular part of the sensory field in order to record from cells responsive to the stimulus, in olfaction we didn't know where to look; we had no a priori knowledge to guide us about where a given odor might be mapped in the olfactory pathway. So although the pioneering studies of Vernon Mountcastle at Johns Hopkins University in the somatosensory cortex and David Hubel and Torsten Wiesel at Harvard Medical School in the visual cortex were our inspiration, their electrophysiological approach to characterizing the mapping of sensory responses in the brain could not be used effectively for revealing the central representation of smell.

The breakthrough came from a visit by Ed Evarts, of the National Institute of Neurological Disease and Stroke in Bethesda, Maryland, to our department at Yale in 1974. Evarts was a leading investigator of the motor cortex. During his visit, I explained our experiments on the olfactory bulb and how difficult it was to know where to look for responses to different odors. Evarts responded that we might be interested in a new method that Frank Sharp, a postdoctoral student in his laboratory, was working on. It was a method being developed in the nearby laboratory of a leading biochemist, Lou Sokoloff. Sokoloff had joined with a great pioneer in the biochemistry of the brain, Seymour Kety, and an outstanding young pharmacologist, Floyd Bloom, to develop a method for mapping activity in the brain based on where the brain uses its energy. The method was based on the fact that nerve cells are exquisitely dependent on oxygen and glucose for their immediate energy needs when they are active; this is why we faint if there is an interruption in blood flow to the brain. The energy is required not for the flow of charged ions across the cell membrane that underlies the impulses or synaptic potentials (electrical changes due to the flow of ions at synaptic connections between nerve cells), but for the membrane pumps that pump back the ions to restore the ionic equilibrium across the membrane.

Sokoloff and his colleagues proposed to track this energy by using a slightly modified form of glucose, an isotope that lacks an oxygen on the number two carbon atom in the molecule; hence the name 2-*deoxyglucose* (2DG). 2DG is taken up like glucose by active cells but cannot be metabolized further. In large doses it therefore blocks metabolic activity, but if given in small doses (called *tracer amounts*) it shows where glucose is taken up without interfering significantly with it. The Sokoloff method involved injecting the substance into an experimental animal, stimulating in the desired way for 45 minutes, and processing the tissue by exposing slices of it to X-ray film to register where the radioactivity was located.

The early results with the method had demonstrated mapping of the visual cortex in the expected way, giving reassurance that the method should give reliable results in mapping brain areas where the results could not be predicted. The significance of this work was far reaching for all of brain science, because development of the 2DG approach led to positron emission tomography (PET) in humans, opening the way to modern brain scans with PET and functional magnetic resonance imaging (fMRI), and their related methods.

The "Aha" Moment

When Ed told us about this new method, it was still being tested physiologically by Frank. Ed cautioned us that one of the possible disadvantages of the method was that it seemed not to label impulse activity but rather mostly activity at the junctions (synapses) between neurons.

It was one of those moments that change one's life. I looked at my postdoctoral colleague, John Kauer, and we realized that this was just what we needed. The incoming fibers (axons) of the receptor cells terminate in the glomeruli at a distance from the mitral cell bodies where the impulse response arises. Electrophysiological recordings from the mitral cell body were therefore a long way from the glomeruli, but the 2DG uptake should be right in the glomeruli where the input from the receptor cells makes its connections to the cells in the olfactory bulb. The 2DG method should therefore be well suited to testing our hypothesis that odors produce spatial patterns of activation of the glomeruli. Evarts and Sharp kindly invited us to join them to test this hypothesis.

The First Smell Patterns

I traveled in December 1974 to the National Institutes of Health to perform the initial experiments. For stimuli, remembering Adrian and the earthworm, I wanted to include something realistic, so Frank and I went to a local supermarket to buy some strong cheddar cheese. An advantage of the 2DG method is that it can be carried out in an animal that is awake, so we put it in a loose holder with its nose into an air stream. We did several controls (meaning with no odor) and several with cheese and with the chemical amyl acetate, which has a fruity banana odor. We did the experiments on rats and rabbits, and, not too hopefully, I left Frank to do the work in preparing the slides.

Early in January, Frank called, elated. There were small dots on the films. Could they mean anything? I asked. Yes, he exclaimed; they were the clearest activity he had seen with the method. Are you sure? I persisted. Yes, he repeated; you should have seen Ed dancing around the laboratory when he saw the dots!

John and I soon returned together to do more experiments, which also gave more positive results. There were sites of increased density on the X-ray films, meaning increased nervous activity, over limited parts of the glomerular layer. So rapidly did the story emerge for us in 1975 that we had to delay our paper to let the paper by the Sokoloff group on the 2DG method come out first. In our first report by Frank Sharp, John Kauer, and me, we wrote:

It appears that there may be a specific pattern of metabolically active sites within the bulb associated with odor stimulation by amyl acetate. This implies that a specific topographical pattern of neuronal activity might be associated with processing of this odor information. Preliminary results obtained with other odors (e.g., camphor, cheese) suggest that there may be differential spatial patterns associated with different odors or odor groups.

The idea that spatial patterns might play a role in olfactory processing is not new; it originated with Adrian and has been the subject of various subsequent studies. The present method recommends itself for direct analysis of this question.

We noted the advantages of the method: that it shows activity throughout the entire system (indeed, the entire brain) and does not interfere with the responses (as an electrode may do poking an active cell). In addition, it can be applied to an animal that is awake and behaving, and it can reveal activity in response to very weak stimuli. All these advantages applied as well to PET and the other methods of functional brain scans in humans.

To get an overview of the activity patterns and compare them with one another, in 1979 our lab, led by William Stewart and John Kauer, developed a pattern mapping procedure based on the fact that the olfactory bulb is approximately like a sphere except for the part where it is attached to the brain. We adapted a "projection" used for world maps, with coordinates reflecting longitude and latitude. It is called the *Molweide projection*; you can look it up in any world atlas.

In neuroscience terminology, this produces what are called *flat maps*. Our results showed activity foci within the glomerular layer, characteristic for a given odor. Taken together, the foci are overlapping but different for different odors. We therefore advanced the hypothesis that the odor discrimination postulated by Adrian could be based on discrimination between glomerular spatial activity patterns. The 2DG results also showed that, at the lowest odor intensities (at threshold for human perception), only one or a few foci were present, each focus presumably representing one or a few glomeruli. When odor intensity increased, so did the extent of activated glomeruli. This suggested that the activity patterns could also encode odor intensity.

Given these promising results, it might be supposed that many other laboratories adopted the method to confirm and extend the idea of odor patterns. However, there were several obstacles. The isotope was expensive, which put the method beyond the budgets and resources of most laboratories. The method required tedious histological procedures. And it required the use of radioactivity, which most physiology or psychology laboratories are not set up to deal with.

In the first extension of the method, Leslie Skeen of Delaware worked with Sharp to obtain evidence for pheromone stimulation of the olfactory bulb in the primate. André Holley and his colleagues in Lyon, France, soon adopted the method and supported and extended our results. They pointed out that the different patterns with different odors meant that odor recognition might fall under the general category of "pattern recognition" in the visual system, an essential idea mentioned in the previous chapter, and which has become a central concept for the neural basis of smell perception.

Archives of Smell Images

The basic findings with the 2DG method have been greatly extended by Michael Leon and Brett Johnson at the University of California, Irvine.

74

(Their Web site contains an archive of more than 500 odor images. These and others can be accessed through SenseLab.)

Beginning in the 1990s, a number of new methods have been introduced for analyzing the activity patterns in the olfactory bulb. Many of them are summarized in an article by Fuqiang Xu, Charles Greer, and me published in 2000. (An archive for fMRI patterns can be found on the SenseLab Web site.) Just perusing Leon's or the SenseLab sites gives you the essential impression of how endlessly variable these spatial patterns are, reflecting the endless variety of smell molecules. The brain uses these patterns to create our smell perceptions.

CHAPTER EIGHT

A Smell Is Like a Face

If you have had a brain or body scan for functional magnetic resonance imaging (fMRI), you know that it is done with your body inserted into a large circular magnet. This technology was developed in the 1990s. Like the use of 2-deoxyglucose (2DG), fMRI is based on local changes in blood flow within the brain that are related to the energy demands of locally active nerve cells. The magnets for humans must have wide internal chambers to receive the human body. Initially they were limited to modest magnetic strengths of 1 to 2 Tesla (the unit for measuring magnetic strength), although recently stronger magnets up to 4 Tesla and above have been introduced. The resolution of the recorded pattern is limited to roughly 0.00006 to 0.00012 cubic inch (about 1 to 2 mm^3), about as wide as the end of thick pencil lead, which enables one to see different regions within the brain but not different layers within a region.

A New Approach: Smell Patterns Using fMRI

As fMRI was being developed for humans, the basic research to understand the method was being carried out in parallel in small animals such as rodents. The magnets for rodents have small, narrow chambers only a few inches across and go to higher magnet strengths, currently up to 11 Tesla and beyond, giving a higher resolution down to approximately

100 μm (0.1 mm). This is about the size of an olfactory glomerulus. In addition to the higher resolution, fMRI offers the opportunity to test different smells, use briefer stimuli, and record faster responses.

When I realized the technology could be applied to activity patterns in smell, it was time to act. As it happened, one of the world centers for developing fMRI in animal experiments was at Yale, just two floors down from my laboratory. Charles Greer, my colleague from our 2-deoxyglucose (2DG) work, and I sat down with the director, Robert Shulman, and his colleagues Douglas Rothman and Fahmeed Hyder to outline our proposal, emphasizing how great it would be if we could image the glomerular layer and even identify single active glomeruli. They immediately got excited and showed us an fMRI study they had just finished on a part of the rat brain that contains a representation of the whiskers, in which each of the 24 whiskers on a rat's snout has its own group of cells, called a barrel, in the cortex. They showed us experiments in which they were able to tweak a single whisker and show activity in its corresponding barrel. It didn't take long for us all to be persuaded that the olfactory experiments should work.

Our experiments started with a 4.7 Tesla magnet. Greer and I worked with the imaging group to devise an "olfactometer" (a smell-delivery apparatus) for stimulating a rat lying anesthetized within the narrow chamber of the magnet. The first images exceeded our hopes, showing strong fMRI activity patterns apparently within the glomerular layer just as in the 2DG case. Our colleagues told us that these were the strongest fMRI signals that they had seen in any part of the brain. We could immediately explain that this was likely due to the fact that each glomerulus is the site of convergence of thousands of incoming olfactory nerve fibers, focusing all the responses of those cells on their corresponding glomerulus.

The patterns demonstrated by fMRI in the rat olfactory bulb were consistent with the patterns obtained with 2DG: different odors give different patterns; the patterns have medial and lateral domains; the patterns are similar in the two olfactory bulbs; and the patterns increase in extent with increasing odor concentration. We could now apply this "high-resolution fMRI" method to a key question: Could the smell activity patterns reflect the different responses of olfactory receptors to chemically related odor molecules in the same animal?

Different Receptor Responses Encoded in Different Odor Patterns

To answer this question, we shifted our study to the mouse. The mouse was rapidly becoming the animal of choice for applying gene engineering because of its rapid reproduction cycle. It offered the promise of labeling different olfactory receptor cell populations and relating them to the smell patterns.

The mouse is much smaller than the rat and therefore poses an extreme challenge to the high-resolution fMRI approach. In contrast to an adult rat, which may weigh about 12 ounces (300 grams) or more, a mouse may be only one-tenth the size, around 1 ounce (30 grams). The mouse olfactory bulb is correspondingly much smaller. However, by then our imaging colleagues had moved up to a 7.6 Tesla magnet. By technical wizardry they were able to image down to about $100 \mu m^2$ ($0.0001 mm^2$), which is near the size of a mouse glomerulus.

Fuqiang Xu, a postdoctoral fellow in the laboratory, took responsibility for carrying out these demanding experiments, using the same series of aldehyde molecules—containing a backbone of four to eight carbon atoms—used in the studies of the I7 receptors described in chapter 5. The animals were exposed to odors, with the magnet humming away, and the images were acquired and stored. As with the 2DG experiments, the functional images were overlaid on the anatomical images, and the glomerular activity patterns were reconstructed.

Summary of the Smell Image Evidence

The take-home message is illustrated in figure 8.1. On the left, the smell molecule activates the receptor within a binding pocket. The diagram shows the sheet of receptor cells in the nose. As explained in chapter 7, each cell sends its impulse response through its fiber (axon) to a glomerulus in the olfactory bulb. All the cells containing the same receptor molecule send their fibers usually to a pair of glomeruli on the medial and lateral sides of the olfactory bulb. This means that whenever a particular receptor is activated, the responses of all the cells are focused on these modules. Cells more or less strongly activated cause

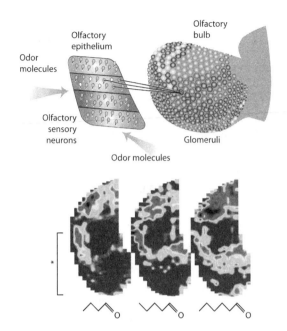

FIGURE 8.1 How the smell image is initially formed

Above, the olfactory receptor cells lie in zones in the olfactory epithelium. Representative cells are shown sending their axons to converge on a glomerulus in the olfactory bulb. The different shades of gray indicate different levels of activity in the glomerular layer. *Below,* flattened maps of patterns of activity in the glomerular layer produced by stimulation with three different odor molecules differing only by a single carbon atom. The bracket indicates the extent of the glomerular pattern on the medial side of the olfactory bulb shown above.

(Adapted from G. M. Shepherd, Smell images and the flavour system in the human brain, *Nature* 444 [2006]: 316–321)

more or less strong activation of their corresponding glomeruli, resulting in a pattern such as that shown in the figure, as seen in this case from the medial surface.

In the lower part of figure 8.1 are three images of the activity patterns in the glomerular sheet constructed from imaging of the functional MRI responses to three closely related aldehyde smell molecules. You can relate the lower part of the image in the left panel (see asterisk and bracket) to the image on the medial side of the olfactory glomerular layer shown in the view of the olfactory bulb, in the upper part of the image. These molecules vary only in their number of carbon atoms, from four to six.

These spatial patterns can be regarded as representing the information carried in the odor molecules. By the analogy with the visual system, where a pattern on the retina is called a *visual image*, we can call a pattern on the olfactory bulb an *odor image* or *smell image*. These results indicate some of the key features of odor images: they extend over much of the olfactory bulb, and they are overlapping but different. As with 2DG patterns, they become more extensive with increasing concentration.

Behavioral Tests

Although these patterns, even while overlapping, are different, can a mouse actually distinguish behaviorally between these small differences? To test for this I was joined by Matthias Laska, then from Munich, Germany. Laska is a leading psychophysicist in the study of smell in monkeys and other species. He and a Yale undergraduate student, Dipa Joshi, carried out experiments in a behavioral olfactometer, consisting of a plastic box and two small openings for delivering two odors. Animals were trained to test the odor at each port and to indicate when they were different, for which they received a small reward. Most of the olfactometer consisted of a complicated array of glass and Teflon tubes and valves to ensure that the odors were pure and that they were delivered in brief pulses.

The results showed that mice are extremely good at distinguishing between these smell molecules. This applies to two odors that are different by only one carbon atom, as well as those that differ by two or more carbon atoms. This discrimination is much finer than shown by the immune system, in which an epitope, as mentioned in chapter 4, consists of up to a dozen or more amino acids within a protein chain on the antigen. By contrast, the ability to discriminate a single carbon atom appears to put the olfactory system in a class by itself.

Many Methods, Many Patterns

Any given method in biology has advantages and limitations. The methods using 2DG and fMRI display smell activity patterns that extend throughout the olfactory bulb, but they use persistent odor stimulation instead of short whiffs, and they can't quite image a single glomerulus. As indicated in chapter 7, investigators have developed many other methods. Some of these involve microscopic viewing of the activation of individual glomeruli. This depends on labeling cells with fluorescent dyes that are sensitive to electrical changes in the cells, or observing minute changes in the microcirculation—so-called intrinsic imaging. Other methods use electrophysiological recordings of nerve cell activity. The results have extended the general conclusions from 2DG and fMRI to more local detailed levels. They show that when applied to the same series of aldehydes, the principles of overlapping but different patterns applies at the level of individual glomeruli.

A disadvantage of these microscopic methods is that one is usually limited to observing only the dorsal part of the olfactory bulb, which may account for only 10 to 15 percent of the surface. It is a bit like looking at a face and seeing only an ear or an eyebrow.

To overcome these limitations, Kensaku Mori and his colleagues in Tokyo in a study reported in 2010 have opened the view to the lateral surface of the olfactory bulb as well. This has permitted much more extensive exploration of the activity patterns. They have been able to plot the responses to a wide range of odors. In general, odors that are related chemically to each other activate glomeruli near each other. There is a tendency toward clustering of glomeruli responding to related types of odors.

With so many patterns from so many methods, it becomes essential to be able to compare them. The field is just beginning to do this. The Sense-Lab database is designed to aid in this effort. As indicated at the end of chapter 7, one can explore the thousands of different receptors, the different odor molecules that interact with them, and the odor maps that are produced by their interaction.

Odor Images, Pattern Recognition, and Faces

All these studies have supported and extended the initial findings obtained with the 2DG method. The patterns have been found in several vertebrate species, including fish, salamanders, mice, rats, rabbits, and monkeys. Odor patterns have also been found in the corresponding parts of the olfactory brains of invertebrates, including honeybees, fruit flies, and the tobacco hornworm. This shows that not only are the glomeruli a constant feature of the architecture of the smell pathway across the animal kingdom, but that activity patterns are a constant feature of their function. Together, these results provide direct experimental proof of the hypothesis that smells are encoded at least in part as spatial activity patterns, and they give the first insight into the mechanism: *the neural basis of odor encoding involves differential activation of the olfactory glomeruli.* When we followed Edgar Adrian's advice and "looked to the glomeruli," we were not disappointed.

This new evidence has stimulated new ways of thinking about the neural basis of smell. For the first time one can say that, just as the nonspatial modality of auditory frequency is represented by a frequency map in the cochlea, which is retained all the way up to the cerebral cortex, so is the nonspatial modality of smell represented in activity patterns, starting in the olfactory bulb. This is what is meant by "the uses of neural space by a nonspatial modality."

Smell Images and Faces

The fact that smell molecules are represented as spatial patterns supports the hypothesis of an analogy with the visual system. When we have a visual perception, we are sensing a spatial pattern of activity that we call a visual image. Likewise, I wish to hypothesize, when we sense an odor we are sensing a spatial pattern of activity that we can also call an "odor image" or a "smell image."

An advantage of thinking in this way is that it encourages drawing on the vast amount of work that has been done, and is being done currently, on pattern recognition in the visual pathway. Because the odor images are irregular patterns of activity, they seem less like the geometrical shapes of many visual objects in our surroundings and more like irregular shapes such as plants and animals, and particularly human faces.

Humans are in fact very good at recognizing faces. The classical illustration of this is that in a room full of grandmothers, you can readily identify your own grandmother. Yet if you are asked to describe your

grandmother's face to someone else, it is very difficult; we lack the vocabulary and an appropriate coordinate system to specify how this pattern recognition is carried out. But we do it unerringly. How?

Recognition of faces is an important subject, not only for designing
machines to perform pattern recognition, but also for law enforcement
agencies trying to identify criminal's faces from witnesses' descriptions.
The process of face recognition has been characterized by Terry Landau
in his book *About Faces: The Evolution of the Human Face*:

> What you see and recognize in [the face of] another person is ulti-
> mately determined by the unique pattern created by the whole picture
> the face presents. That is identity. It's not the details of the features. It's
> not the spacing between features. Rather, it is the interrelationship
> among all of these things taken together [that is, the gestalt] that en-
> ables us to recognize the identity of a criminal or the face of a friend.
> You can't take faces apart analytically. Most of the time you can't even
> tell what you are looking at when you look at a face because verbal
> processes have little to do with visual recognition. Identity is encoded
> on the face and remembered as a whole. Though elusive and hard to
> define, it is experienced on many different levels, not the least of which
> is that identity enables us to carry out an important social skill: recog-
> nizing one another as individuals.

It seems reasonable to hypothesize that recognition of smell patterns
in the olfactory bulb involves similar principles. We see a pattern as a
whole, we learn to identify it with its "smell object," and we learn to
discriminate it from patterns for other smell objects that may be very
similar but carry a different behavioral meaning.

To pursue the analogy a step further, the quotation from Landau applies only to recognizing a face that one has seen only once, in which one
needs the entire gestalt to remember it (barring some noticeable single
feature). This is quite different from recognizing a familiar face one has
seen many times, in which case catching only a fleeting glimpse of part
of the face, or of the face in poor light, can be sufficient; animals are
very good at recognizing these minimal cues. An example is an increasing pixelation of photographs of Abraham Lincoln or the Mona Lisa.
We have no trouble identifying the picture even though the resolution is

poor, because we have trained our pattern recognition abilities on it so that we can match it with our stored memory despite the indistinct image. The adaptive value of this ability is obvious; identifying prey or a predator in the indistinct shadows of dusk is critical to survival, as is the ability to identify the face of a friend or enemy in shadows. Because most animals live by their sense of smell, an analogous ability to identify an odor image would be critical to survival.

In the case of the visual system, it is easy to grasp the concept of a visual image in the retina, because this image is maintained through several steps of processing from the retina to the higher centers in the cortex. By contrast, in the olfactory system, we are not conscious of the fact that our olfactory bulbs have constructed an image of an odor. We may speculate that this is because that image does not represent the real odor world, but is only *the brain's way of representing the odor world*. What the brain cares about is how that image is processed to provide the basis for our perception of different smells and different smell objects.

It is therefore to higher stages of processing in the brain that one must look for understanding how the images are used as a basis for smell perception.

CHAPTER NINE

Pointillist Images of Smell

The neural basis of our ability to perceive a rich palette of smells can be compared with the neural basis of our ability to perceive a rich palette of colors. The best way to illustrate this is by comparison with pointillist art.

There are two ways to put colored paint on a canvas to elicit the perception of color in the mind of the observer. One is to mix the paints to achieve a particular color impression: red and white to achieve pink, blue and yellow to achieve green, and so on.

The other is to place different colors in individual small "points" of color and let the effect of color mixing arise from a distance, where the colors blend to produce the mixed color in the mind of the observer. This was the method of "pointillism." It was invented and perfected by Georges Seurat, who had drawn his inspiration and methodology from visual theorists, most notably Wilhelm Helmholtz. Paul Signac was another well-known practitioner of this art form.

Seurat's painting *A Sunday on La Grande Jatte* (1884), for example, is produced by thousands of tiny dots depicting several dozen figures on a bank of the Seine outside Paris on a sunny Sunday afternoon. The closer one is to it, the more one can see the individual dots of color, but the less one can see the figures they are forming; further away, the dots fuse and the entire picture comes into focus. The painting hangs in the Art Institute of Chicago. (You can appreciate the description that follows by observing the painting in full color on the Web.)

Reproduction by dots in fact became the method of publishing photographs and other graphic art in newspapers and magazines, in which a picture is made of tiny dots, each of a different shade of gray or of color. In art, pointillism reached its reductio ad absurdum in the work of Andy Warhol, who made large dots themselves the object of the artwork in the pictures of cartoon figures and Hollywood celebrities.

Several years ago, Terry Acree at Cornell University and I independently realized that these examples from the modality of vision can serve as useful suggestions of how smells are perceived. The illuminated dots of colored paint reflecting different wavelengths of light are analogous to the modules called *glomeruli* in the olfactory bulb, each activated preferentially and differentially by different odor. The patterns of colored dots seen in *La Grand Jatte* correspond to the patterns of activated glomeruli (see figure 8.1). The perception of the color patterns requires their registering from a distance in order to yield the complex effects of color mixing. By analogy, the perception of smell would require a "distance" provided by a readout that can produce "odor mixing" of the effects of neighboring modules. That distance illusion merges the pointillistic representations of the color patterns, as shown by the patterns in figure 8.1.

The neural circuits in the olfactory pathway in the brain are organized to carry out this series of operations. This series was summarized in figure 7.1. We take up each step along the way to gain insight into how pointillist representations of smells create their perceptual worlds in our minds.

The analogy with color has another very revealing similarity. What we call *color* is a perception that arises from electromagnetic waves of different wavelengths activating our photoreceptors to different extents; it is the brain that creates "color" by the way it processes the signals according to the different wavelengths. Similarly, "smell" is not present in the molecules that stimulate the smell receptors. It is the brain that creates the perception of "smell" from the differences between the features of the different smell molecules. This is a fundamental basis of smell and flavor. Hence, the subtitle of this book. As we move into the brain, we pass from the domain of the stimulus through processing circuits to the domain of the brain that creates our subjective world. Those processing circuits are next.

Comparing Color and Smell

We have pointed out that color is not contained in the wavelengths or photons of light. It is a sensory quality, created by our brains. Every time we view a flower, we are reminded how dull a scene is if the different wavelengths give it only shades of gray instead of, say, a bright red, which is red only because our brains have neural circuits that select that wavelength of light and create that internal perceptual quality. In the same way, we postulate that our smell world would also be only a smear of shades of inchoate sensations set up in our noses if we did not have the circuits in our smell pathway that select specific types of molecule-stimulated activity to give them a quality we can identify as distinct from all the rest. The ability to create this quality—philosophers call it a *qualia*—of a specific smell starts with the remarkable glomerulus.

Processing the Pointillist Image

In all other sensory systems, the brain employs successive stages of processing to extract what is most important for the immediate needs of detecting and discriminating among its stimuli. In the smell pathway, a sequence of stages similarly processes the initial "odor image." Comparisons such as that in figure 7.1 with the visual system make this sequence more intuitive.

First is the processing of the initial odor image in the olfactory bulb at the layer of the modules forming the image. This image is then enhanced by a powerful system of lateral inhibitory microcircuits. The enhanced image is sent to the olfactory cortex, where a cortical microcircuit with widespread connections reformats it into a content-addressable memory. The final stage is for this memory representation to be sent to the highest centers in the neocortex, where a complex cortical microcircuit gives rise to conscious perception.

In sum: form a pointillist image, process it locally, format it globally, represent it in memory, enhance it with emotion, and perceive it consciously.

Each of these steps is performed by its own microcircuit, which will be described in subsequent chapters.

The Glomerulus: A Universal Smell-Detecting Device

In our analogy with pointillist painting, the points are the olfactory glomeruli. The glomeruli constitute an amazing little structure, the most distinct multicellular unit in the brain. Little wonder that Edgar Adrian advised me to "look to the glomeruli," that we have regarded it as the basic functional unit of this sensory system, and that molecular biologists are focused on its organization and function.

A glomerulus is essentially a densely packed meeting place, where signals from the nose are transferred to the brain (see figure 7.1). Here the terminals of fibers (axons) from the receptor cells in the nose connect to the short branches (dendrites) of nerve cells in the olfactory bulb. A single glomerulus receives not just hundreds but thousands of incoming nerve fibers from the receptor cells. In the rabbit, for example, there are some 50 million olfactory receptor cells and some 2,000 glomeruli. This gives a ratio of, on average, 25,000 cells for each glomerulus. We say that this defines a *convergence ratio* of 25,000:1. This is one of the highest ratios of a given cell type and a given target in the brain.

The molecular biologists Linda Buck, Richard Axel, Peter Mombaerts, and their colleagues have shown that all the fibers coming to one glomerulus express only one type of olfactory receptor in their cilia in the nose. This means that all the fibers are carrying the same information, making the convergence ratio even more impressive.

Imagine that 25,000 people are talking to you at the same time. How would this work? It would not if it involves being in a monster cocktail party where everyone is saying something different or the same thing at different times. We would call this *noise*, and technically this is what it is. However, if they are all shouting, in unison, "Happy Birthday" or singing a song together, the words will be loud and clear. Technically, this is the "signal." In signal theory, the converging simultaneous inputs raise the "signal-to-noise ratio," thus making the signal much more distinct.

We postulate that *signal-to-noise enhancement* is a key operation of the glomerulus. This is used in orthonasal smell to detect and discriminate among specific signals in the environment that may be critical for survival, such as the scent of prey to attack or predators to avoid. Similarly, it is likely used in retronasal smell to detect and discriminate the volatile components of food released within the mouth. These components may also be critical for survival, in signaling the ripeness of fruit or whether fish or meats are fresh or rotting. They may have been critical for exploring new food sources, as during the great migrations of people that occurred during human evolution. And today, we use them to enjoy our food and discriminate among our wines.

Why should nearly all animals, including humans, need this module for their sense of smell? The odor environment consists of thousands of different odor molecules and odor objects, out of which we must identify only those that are behaviorally significant. This environment constitutes the background of asynchronous noise, out of which the animal must detect the signal. To do this, the olfactory system does not have to spend expensive neural tissue to track its stimuli in space, as does the visual system with its constantly changing visual scenes. Instead, the olfactory system can build immovable detection modules that wait for their appropriate stimuli to come to them. To adapt the analogy made famous by the Oxford philosopher Isaiah Berlin, the visual system is like the fox, in which each location knows many things, and the olfactory system is like the hedgehog, in which each glomerulus knows one thing.

The glomerulus is thus a detection device par excellence. Just as the points of paint at a particular wavelength in a pointillist painting give brighter colors than if the paints were mixed on the palette, so we can imagine that the points of "smell molecule features" represented by a glomerulus shine more brightly (can be perceived more sharply) against a mixed-smell background. Note how these follow the principles of pixelation of a visual image explained in the previous chapter. This special property may be an important reason why the glomerulus is the nearly universal module in the processing of odors by animal species, from humans to insects.

How Fine An Image?

In our analogy with vision, the fineness of resolution of the odor image must depend on several factors. First, of course, is the number of receptor cells; presumably the more the better. This is consistent with the dog having up to 100 million, around 10 times the number for rodents and humans. The next factor is the number of different types of smell receptors. Here the rodent comes out on top, with over 1,000, while the dog has some 800, and the human has around 350. Finally is the number of glomeruli. The dog has several thousand glomeruli, and my colleague Charles Greer and his colleagues have recently shown that the human has even more, up to around 6,000. It appears that more receptor cells, receptor types, and glomeruli give higher resolution in our smell images. However, the greater complexity of the human brain in analyzing the images becomes a critical factor, as we shall see.

Single Modules Interact Specifically

So far in the human, we have several thousand modules, each responding independently. They are acting like separate "labeled lines." But it is another axiom of signal theory that the receiver—the brain—cannot make any sense of the information in those labeled lines until it can compare them. There must be mechanisms for correlating and comparing the activity of the pathways with one another. Specifically, lateral interactions between the glomerular modules are needed. These interactions begin through interneurons called, appropriately, *periglomerular cells*, which connect to neighboring glomeruli (see figure 7.1).

Physiological studies have shown that the periglomerular cells respond to odor input with single impulses or with impulse bursts. One of the physiological actions that has been identified is inhibition of mitral and tufted cell dendrites issuing from surrounding glomeruli. It has been hypothesized that this may enable a more active glomerulus to inhibit its less active neighbors—the kind of lateral inhibition we have discussed. This action would contribute to signal-to-noise enhancement of the odor image.

Another kind of interglomerular action may be excitation. This could involve direct excitatory synapses from a subtype of periglomerular cells that has an excitatory neurotransmitter. But it could also involve a special type of inhibition, in which an excitatory periglomerular cell axon excites a distant inhibitory cell, producing a distant but local inhibition of the mitral and tufted cells in the target glomerulus. In a study conducted in 2003, Michael Shipley and his colleagues at the University of Maryland obtained evidence for this kind of action. Some of the periglomerular cells with these actions target glomeruli at considerable distances away, indicating that extraction of the odor pattern involves complex interactions that are coordinated over several glomeruli. Tom Cleland and his colleagues at Cornell University have suggested that the global effect of these actions may be to normalize the excitability at the glomerular layer, to keep it within a working range, regardless of the intensity of odor stimulation.

We are only starting to learn about the subtypes of periglomerular cells and the kinds of interactions that occur between glomeruli. At this point it seems safe to say that one effect of these lateral interactions is to begin the extraction of the spatial pattern so that it can be read more effectively by the next level of microcircuits involving one of the brain's most enigmatic cells, the olfactory granule cell.

CHAPTER TEN

Enhancing the Image

There are two levels of processing in the olfactory bulb. The first, described in chapter 9, consists of the glomerular layer, which forms an image representing the smell molecules and performs signal-to-noise operations and lateral interactions to begin the processing of the image. The image is then sent to the second level within the olfactory bulb. The connection between levels is by means of a large cell, the mitral cell, and its smaller companion, called a *tufted cell* (see figure 7.1). These cells collect the input in their dendritic branches in the glomerulus, transfer the processed signal to the next level through a long primary dendrite, and send it out through their long axon to the olfactory cortex. But before it goes out, the image is subjected to an additional stage of processing. This is necessary because the glomerular interneurons have had their actions on the odor image limited by their axon connections to specific surrounding glomeruli.

At least two operations are required before the image can be sent further. First, there must be coordination with all other glomerular modules. This takes place through long so-called secondary dendrites of the mitral and tufted cells, which reach out sideways for long distances, branching and terminating across many neighboring glomerular modules. These dendrites do not interact with each other, but rather with the special type of interneuron called a *granule cell* (see figure 7.1). Through it, the second operation—lateral inhibition between coordinated glomerular modules—is carried out. There are about a hundred granule cells for every mitral cell, so this is strong inhibition. The granule cell is thus a key to this coordinated inhibitory processing, whose function is to format the odor image for output to the next stage, the olfactory cortex. How does it do it?

Solving the Granule Cell Problem

I first encountered the granule cell as a graduate student with Charles Phillips in Oxford when I was studying the physiological responses of cells in the olfactory bulb. My main finding, simultaneous with the same finding in two other laboratories, was that mitral cells are subjected to very strong and long-lasting lateral inhibition.

But there were no known inhibitory cells in the olfactory bulb at the level of the mitral cell secondary dendrites, only this curious little granule cell with prickly looking spine-covered dendrites, and no axon. Without an axon one could not even be certain that it was a nerve cell, much less one that could be an inhibitory interneuron for the mitral cells. Nonetheless, it seemed to be in the right place, so we suggested that it could be activated by side branches (collaterals) of the mitral cell axon and inhibit the mitral cell through its long dendrite mingled among the secondary dendrites.

The next step in my training consisted of postdoctoral studies with Wilfrid Rall at the National Institutes of Health (NIH) in Bethesda, Maryland. Rall was establishing himself as the pioneer of computational neuroscience, constructing the first computer models of nerve cells to elucidate the functions of their enigmatic dendritic branches. He had to endure much opposition from those who believed that the dendrites were not involved in information processing but were there primarily for nutritional support. We decided that the mitral and granule cells of the olfactory bulb were excellent subjects for his approach to support his theory of the importance of dendrites and my theory that they were important for processing smells.

Our aim was to build computational models of the mitral cell and the granule cell to test whether we could account for my experimental recordings. Almost two years of work produced a model for impulse spread in the mitral cell dendrites, excitation of the granule cells, and inhibition of the mitral cells. Unfortunately the model didn't give any new insights into how the two kinds of cells might actually interact. I had another couple of months before leaving for my next position. The inhibitory action of the granule cell on the mitral cell seemed clear enough, but how was it activated to begin with? The more we struggled with the problem, the more the constraints of the model indicated that the excitation of the granule cell dendrites must occur at the same narrow level as the subsequent inhibition of mitral cell dendrites. But how?

More "Aha!" Moments

As we discussed this problem one afternoon, the idea hit us that the excitation of the granule cell dendrites must come from the same mitral cell dendrites that the granule cell dendrites then inhibit. It sounded crazy, but it seemed the only solution. I knew from the classical anatomical literature that there was no precedent for this kind of interconnection between dendrites, and we both knew there was no precedent in the physiological literature for this kind of functional interaction. So Rall duly wrote down our hypothesis in his green protocol book (on August 26, 1964, to be exact): "dendrodendritic interactions" between mitral and granule cells likely occur, and these would function to cause self and lateral inhibition of the mitral cells. We suggested that this would be similar to the lateral inhibition in the retina. I then went on to a further research fellowship in Sweden.

The only sure way to test this hypothesis was with the electron microscope, which would enable one to prove that such synapses existed. As it happened, Tom Reese and Milton Brightman were working at NIH in a nearby building. I had encouraged them to look for synapses between the mitral and granule cell dendrites, and they soon found them. On being shown that the synapses were unusual in being situated side by side and oriented in opposite directions, Rall had another "aha!" moment. He immediately told Reese and Brightman that these were precisely the kinds of contacts to mediate the interactions we had postulated. It was an example of how perplexing data can be interpreted instantly by the prepared mind.

My next "aha!" moment came in Stockholm on receiving the letter with this news. Two other investigators reported the "atypical configurations" in the olfactory bulb, but without the physiology and the model it wasn't possible to infer a function for the synapses because they seemed to be in direct opposition to each other.

Shaping Spatial and Temporal Processing

The four of us excitedly wrote up our initial results and submitted them to the leading journal *Science*. The reviews came back: "Rejected; not of general interest." We could have fought against the rejection, as is the custom these days, but Wil was too courteous for that. He found, instead, another journal that would take our paper.

We subsequently published the full details of the model in a second paper. This made explicit the way the self-inhibitory interactions always proceed in a sequence from excitation of a granule cell to feedback inhibition, so that there is no opposition, whereas lateral inhibition is mediated only by the granule-to-mitral synapse onto a neighboring mitral cell. This provided a new type of circuit, for mediating more localized self and lateral inhibition compared with the classical pathway. We pointed out that the lateral inhibition was likely to be important in the spatial organization of activity in the olfactory bulb as well as contribute to oscillatory activity in the mitral-granule cell populations. Thus, the same interactions are at the core of both the spatial and the temporal properties involved in processing the odor images.

Later, in an article for *Scientific American*, I termed these and other interactions like them *microcircuits*, in analogy with microcircuits in computers. The term has caught on and become useful in describing the organization of the nervous system in terms of distinct and repeatable patterns of connections, for which the olfactory bulb is still one of the best models. It has become an example of how a theoretical model can predict a circuit with a specific function. After more than 40 years of testing, it continues to be helpful in guiding experiments on the microcircuits for odor processing in the olfactory bulb.

How Multiple Glomerular Units Are Coordinated

How are the widely distributed glomerular modules coordinated in processing the image? We have seen that the process begins in the localized connections of periglomerular cells between glomeruli. We now need to see how it is completed by the lateral inhibitory connections through the granule cells.

The problem is this: the odor image extends broadly within the olfactory glomerular layer, even when aroused by a single odor molecule (chapter 8). It requires that the responses of mitral and tufted cells that may be far apart need to be coordinated so that lateral inhibition can occur to enhance the image. How do we get effective lateral inhibition over long as well as short distances?

An answer to that question has come from a new method based on the virus that causes rabies. As is well known, the rabies virus enters nerve cells and is transported throughout their branches, killing them, and also passing into cells to which they are connected in order to kill them as well. Molecular biologists have taken advantage of this property and turned it into a research tool for understanding how nerve cells are connected—by modifying the virus into a "pseudorabies" virus so that it has a much reduced lethality. By attaching a fluorescent label to this virus, its progress through the cell and into other cells can be traced.

David Willhite, a postdoctoral student in our laboratory, used this tool to trace the connections between mitral and granule cells. This showed that an infected mitral cell is connected to a local cluster of granule cells, mitral cells, tufted cells, and periglomerular cells that are all related to a single glomerulus. The cells form a column, somewhat similar to columns of cells that have been seen in areas of the cerebral cortex, for processing the glomerular signal. We have termed this a *glomerular unit*. Willhite finds that a given mitral cell has connections to other glomerular units that are widely distributed through the olfactory bulb. It appears that this could be the basis for the hypothesized coordinated lateral inhibition.

How can one mitral cell impose strong lateral inhibition on mitral cells of different glomerular units over variable distances between them? A key was provided by two other colleagues, Wen Hui Xiong and Wei Chen, who showed that the impulse generated in the mitral cell body not only travels out into the axon and on to the olfactory cortex, but also travels backward into the lateral dendrites all the way to their tips. This means that the dendrite can act somewhat like an axon in spreading impulses long distances to activate granule cells and their lateral inhibition throughout the olfactory bulb.

Another colleague, Michele Migliore, constructed a computational model in which the impulse activates granule cells belonging to glomerular units at variable distances, thus providing the strong inhibition at arbitrary distances required for processing the distributed odor images. It was the first major change to our original model in 40 years. And just as with the first model, it was rejected as being "not of general significance" the first time around, before eventually being accepted by another journal.

These inhibitory operations not only provided for enhancement of the spatial images, but also for synchronization of the activated mitral cells through inhibitory gating of the mitral cell responses, as predicted in the original model (the odor pattern is therefore one that exists in both space and time). There is much to test in this new proposal, but it provides a working hypothesis that extends the original model and can guide experiments in the future.

The reader will note that this new view has come about from the coordinated research of several groups. This illustrates several important features of doing basic research: the need for stable funding over a sufficient period of time; the need for a critical mass of investigators working on different aspects of the same problem; and the advantage of coordinating computational with experimental approaches. These were behind our success in the original discovery of the dendrodendritic synapses and their mechanisms, and they were behind the recent advance in understanding their coordinated action among the distributed glomerular units.

Modulating Our Appetites

In addition to providing for this important step in processing the sensory input, the granule cells are also a key site of modulation by the behavioral state of the animal. By *behavioral state*, we mean whether the animal is awake or sleeping, or whether it is hungry or satiated. The olfactory bulb is obviously an important step with respect to the brain mechanisms for feeding; if we are hungry, the smell of food (either the aroma or the flavor) really stimulates our appetite; when we are full, the smell is much less attractive, perhaps even aversive. This modulation is obviously critical for the brain flavor system. It turns out that processing in the olfactory pathway is heavily dependent on whether we are hungry or full, and that this starts in the olfactory bulb itself.

This was first shown in 1978 by a French scientist, Jeanne Pager, who recorded from mitral cells in rats under these two conditions and found that they showed brisk responses if the animals were hungry and weak responses if they had eaten. The most likely interpretation is that this reflects the activity of central sites in the brain that are influenced by feeding signals, such as blood sugar and intestinal sensory fibers. The granule cells receive heavy inputs from these central sites (particularly the site known as the *locus ceruleus* in the brain stem and the *diagonal band* in the forebrain). These outwardly directed "centrifugal" fibers are indicated in figure 7.1, coming from the brain stem and from the nucleus of the horizontal limb of the diagonal band (NHLDB) in the forebrain. It has been hypothesized that activity in these fibers functions as a set point for gating the flow of sensory information about smell through the olfactory pathway in relation to the feeding status of the animal. They are increasingly busy modulating retronasal smell patterns through the granule cells and the glomeruli as you proceed through your dinner and go from feeling hungry to feeling full.

Similar modulatory actions take place at every stage deeper in the brain. They make the olfactory pathway probably the most heavily modulated pathway in the brain. The reason for this appears to be that our perception of food smells is heavily dependent on our behavioral state: whether we are hungry or full, angry or sad, craving for something or repulsed by it, suspicious of a new food (called *neophobia*) or eager for novelty (*neophilia*). The smell image has to be modified by the behavioral state. The next chapter will discuss further how appetite modulates a central brain region involved in flavor.

CHAPTER ELEVEN

Creating, Learning, and Remembering Smell

In all mammals, the output fibers from the olfactory bulb gather in a bundle called the *lateral olfactory tract*, which connects to the next stage, the olfactory cortex. The tract is relatively short in most animals, but very long in humans, an inch or so (up to 30 mm), in order to reach from the olfactory bulbs, sitting in front over the nasal cavity, to the olfactory cortex on the underside of the brain. The length reflects the expansion of the brain as the neocortex grew in size during mammalian and especially primate evolution. As in the case of the olfactory nerves, all our sensations of smell depend on this connection.

The olfactory cortex is little noticed when the many higher cognitive functions of the cerebral cortex are discussed. What is the olfactory cortex for? Why is it that the coordinated, multidimensional smell image from the olfactory bulb cannot be sent straight to the highest cortical level—the neocortex—to serve as the basis of perception?

Modern research reveals the olfactory cortex to have remarkable properties that make it an essential player in the human brain flavor system. Most significantly, it represents the transition from the steps of extracting features in the smell stimulus to the steps of creating the perceptual qualities of smell. It is where the external features of the outside world meet the internal features of our perceptual world.

If you are able to smell something, it is because of the olfactory cortex. Similarly, if you sense the flavor of something, it is because of the olfactory cortex. Understanding how the olfactory cortex functions has to be an important subject in neurogastronomy.

Out of the extracted odor image of the stimulus in the olfactory bulb the olfactory cortex creates the basis for the human perception of a unified smell perception, what is called an *odor object*. How does it do it?

Introducing the CEO of the Cortex

To answer this question, we look inside the olfactory cortex to see what happens to the odor image. These experiments can't be carried out in the human or the monkey, so the laboratory rat and mouse are necessary for studying them.

The main nerve cell in the olfactory cortex is a *pyramidal cell*. As shown in figure 7.1, its cell body is shaped like a small pyramid, giving rise to a large "apical" dendrite that ascends to the surface and several "basal" dendrites from the base of the pyramid. There must be something very useful in this arrangement of dendrites, because the pyramidal cell is the main kind of cell in all types of cerebral cortex. One could call it the CEO, the Chief Executive Officer, of the cortex. It is worth knowing something about this kind of cell. Your mind depends on it for normal thinking, and it is the target for degeneration in Alzheimer's disease. You can say that your mind is what the pyramidal cell does.

To pursue our analogy, this CEO has two possible actions, which it exerts through branches of its axon. One is to send impulses to excite its co-workers, the interneurons. As in the olfactory bulb, and in fact as in most parts of the brain, the interneurons feed back inhibition onto an excited pyramidal cell, controlling its output, and onto neighboring pyramidal cells, to sharpen contrast. So far this seems similar to the way a mitral cell is organized.

However, there is a big difference that accounts for much of what goes on in all areas of the cortex. The axon collaterals also feed back excitation onto an excited pyramidal cell and its neighbors. These re-excitatory feedback collaterals in the olfactory cortex were discovered in 1973 by Lewis Haberly, using physiological recordings when he was a graduate student in my laboratory at Yale. They were also independently revealed, using anatomical methods, by Joseph Price at Washington University in St. Louis.

Feeding back excitation onto an already excited cell seems like a recipe for runaway excitation, but in fact it is a fundamental element in not only the olfactory cortex but in all types of cerebral cortex. In normal function this feedback excitation is counterbalanced by the feedback inhibition through inhibitory interneurons. We suggested that together with the feedback inhibition this forms a "basic circuit" for all cortical regions, and that the basic circuit is modified in the different cortical regions for the specific functions in each. This basic circuit is similar to the "canonical circuit" for the cerebral cortex, suggested by Rodney Douglas and Kevan Martin, then at Oxford University.

The Olfactory Cortex Creates a "Content Addressable Memory"

The fibers in the lateral olfactory tract carrying the odor image pass along the surface of the olfactory cortex and send many branches to the underlying layer, where they transfer their activity by means of synapses onto the most distal apical branches of pyramidal cells (see figure 7.1). There is thus a radical difference from the way input comes into the olfactory bulb, where receptor cells with the same response sensitivities—that is, the same "molecular receptive range"—converge onto one glomerular module. The mitral cells, the output of which still reflects the glomerular module to which they are connected, thus distribute their output across the olfactory cortex to many pyramidal cells. In this way, the information is changed from a mosaic image to a distributed representation of that image.

What is the nature of this distributed image? An answer was suggested by Haberly. After finishing his dissertation, he took postdoctoral training with Price in St. Louis, bringing together the two people who had put the re-excitatory collaterals on the map. Starting with the evidence for the smell patterns being sent to the olfactory cortex from the olfactory bulb, Haberly began to study the literature that was emerging on pattern recognition devices. He compared the anatomical and functional organization of the olfactory cortex that he knew so well with other parts of the brain doing similar kinds of pattern recognition.

In 1985, Haberly wrote a classic paper on the olfactory cortex in which he suggested that the "olfactory cortex serves as a content-addressable memory for association of odor stimuli with memory traces of previous odor stimuli." He noted the properties necessary for the cortex to function in this manner: a large number of integrative units (pyramidal neurons and synapses) relative to the number of memory traces; a highly distributed, converging-diverging, input (from the olfactory bulb fibers); and positive feedback (the re-excitatory axon collaterals) via highly distributed interconnections between units. All these properties are found in the basic olfactory cortical microcircuit. Not surprisingly, this basic microcircuit is similar to that in the hippocampus, which is well known for its role in long-term memory (chapter 21).

The final property necessary for a cortex to mediate learning and memory consists of the synapses that are reinforced by coincident action of presynaptic and postsynaptic activity. This is the so-called Hebb rule, named after the psychologist Donald Hebb, who in 1949 suggested that this coincident action would build memory into brain circuits. This too is a property of the synapses in both olfactory and hippocampal cortices.

Haberly was particularly intrigued by a comparison of the recognition of odor images with recognition of faces in the visual system. He noted that processing of the highly distributed and complex patterns of activity in the olfactory cortex is different from the initial feature extraction known to be carried out in the primary visual cortex, resembling more closely the discrimination of complex visual patterns such as faces carried out by higher-order visual association areas. We have already seen the usefulness of a comparison with the visual system. As discussed in chapter 8, there is thus an analogy between recognition of the complex patterns laid down by odors in the olfactory cortex and recognition of the complex patterns of faces in visual association areas. By studying the microcircuits, we can now see that in both cases, the extensive horizontal connections through re-excitatory collaterals are essential for the storage and recognition mechanisms.

Olfactory Cortex Matches Inputs to Memory

This has been a fertile hypothesis for the field of smell. An elegant summary of the experiments supporting it is contained in *Learning to Smell: Olfactory Perception from Neurobiology to Behavior* by Donald Wilson of New York University and Richard Stevenson of McQuarrie University in Sydney, Australia. The main point is that, whereas the representation of smells in the olfactory bulb is driven by stimulus properties, the representation in the olfactory cortex is memory based. Wilson and Stevenson identify several defining characteristics of how this works, which may be summarized as follows:

1. The olfactory cortex responds especially to *changes* in its input signals from the olfactory bulb; it adapts (reduces its responses) to continued stimulation with the same smells. This, in fact, was the reason that first attempts using functional imaging in the human failed to identify responses to smell in the olfactory cortex. Noam Sobel and his colleagues realized that the responses were adapting out, and used changing smells to reveal the olfactory cortex activity.

2. This system *learns*. The basic cortical circuit has the ability to improve its performance with repeated exposure to different smells. The *recurrent excitation* strengthens the cells activated by input from the olfactory bulb. The *lateral inhibition* enhances the contrast between activated and less activated cells. Finally, the *synaptic strengths* change so that the system can store these changes as a memory and match them to the input.

3. These changes with learning enable the system to improve its ability to *match* an input pattern to a stored pattern, so that finer discrimination between more similar smell molecules can occur.

4. These changes also enable the system to improve its *signal-to-noise ratio*, so that detection and discrimination of a particular smell can be enhanced against a background of many smells.

5. The olfactory cortex microcircuit functions to take an input reflecting many diverse stimuli and construct out of it a *coherent odor object*. This is an analogy with how the visual system takes an input of different shapes and sizes and constructs a "visual object." An important aspect of such an object is that seeing only a small part of it still enables the system to "fill in" the missing parts. This is a feature of all sensory experience. As we have noted, we catch a glimpse of a distant figure, a few notes of a melody, a whiff of a smell, and can instantly "fill in" the rest. This especially applies to perceiving a face or just a part of a face, as discussed in chapter 8. It also means that the system "degrades gracefully" if damaged.

6. The odor object is in a form that can be *combined* with other sensory inputs to produce the sensation of flavor. This occurs at the final level in the smell pathway, the orbitofrontal cortex.

Where Does Conscious Smell Perception Arise?

Given these impressive properties of the olfactory cortex, a key question is whether conscious perception of smell arises there. I have asked many behavioral psychologists this question, but apparently the crucial experiment of interfering with the next stage has never been done, at least not in primates. (For evidence from trauma to that level in humans, see chapter 25.)

The argument in favor of conscious perception arising in the olfactory cortex is that, although many call this the primary olfactory cortex, it is not really equivalent to the primary cortex in other sensory systems, where that term is reserved for the receiving area in the neocortex. The olfactory cortex, as Haberly suggested, is more equivalent to a higher-association area in other sensory systems, even though it is not yet at the neocortical level.

The argument against conscious perception arising in the olfactory cortex is that the smell information has not yet passed through the thalamus or reached the level of the neocortex, which is necessary for all other sensory systems.

So we need to go to the next level—the orbitofrontal area of the neocortex—for possible answers to where conscious smell arises.

Detecting Essential Amino Acids

In addition to the role of the olfactory cortex in smell perception, another function began to emerge some two decades ago with the evidence that it contains an area sensitive to amino acids in the diet. Of the 20 amino acids required for building proteins, 10 are essential; if one of them is missing from the diet, an animal's health begins to fail and it will die if the deficit is uncorrected. Rats will cease feeding within 30 minutes if their chow lacks just one of these. What is the mechanism?

A series of studies over the past 20 years has shown that the sensor is in the brain and, surprisingly, has narrowed it to the olfactory cortex. Evidence from Dorothy Gietzen at the University of California, Davis, and her colleagues Shuzhen Hao and Tracy Anthony published in 2007 indicates that the pyramidal cells contain a molecular mechanism that senses the lack of an essential amino acid by its inability to "charge" its appropriate transfer ribonucleic acid (tRNA) molecule with that amino acid. How this is communicated to the cell membrane to change the cell's activity, and what is the pathway for communicating this message to the rest of the brain, is still under study. One can speculate that the exquisite balance between excitation and inhibition in this region enables the cells to be sensitive detectors of slight changes in the presence of the amino acid. This is an unexpected hidden function of the mammalian olfactory system, one that is apparently present in humans. It may be crucial to the nutrition of people in conditions of poverty and starvation around the world. It may also be a critical element in the need for vegetarian diets to include foods with all the essential amino acids. The relation to flavor is only indirect, but it presumably means that for an omnivore such as ourselves vegetable flavors must be learned in order to supply the needed amino acids.

PART III

Creating Flavor

CHAPTER TWELVE

Smell and Flavor

We have seen that the olfactory cortex begins the task of creating the brain's representation, not of individual smell molecules, but of "smell objects" representing the food we are consuming. What more is needed? The main need is someone to "read" the smell object in a way that gives it meaning in a human context. This requires the neocortex, in evolutionary terms the newest type of cortex, that dominates the mammalian brain.

The increase in the area of the neocortex in primate and human evolution has produced many different cortical areas. These areas are of three main types. First are the areas that connect with the sensory and motor pathways below. These are called the *primary sensory and motor areas*. They are relatively large in the human in order to carry out the initial processing of sensory input and the final control of movements at the neocortical level. Second are the *association areas*, specializing in elaborating the properties of a given sense—for example, in vision there are areas for elaborating the processing of signals about color, movement, and faces—and coordinating motor acts. Finally, there are *higher association areas*, which are concerned with creating our highest mental faculties. These include language—both the areas related to interpreting language and those related to producing speech—and higher cognitive functions such as reasoning and planning ahead. These areas multiply the representations of our sensory and motor worlds. They also multiply the number of connections between areas. In addition, each area generates internal states that further abstract those worlds. The attributes that we consider human depended on a great expansion of these cortical areas during human evolution.

The smell image fashioned in the olfactory cortex has a privileged place in this expansion.

Neocortical Smell Perception

To get to the neocortex, the pathways subserving other senses—vision, hearing, touch, and taste—pass through the thalamus, often called the *gateway to the neocortex*. The thalamus sends forward and the cortical areas answer back, so that they function together in a coordinated manner. These other sensory pathways are located in the middle and back of the brain.

For smell it is different. The olfactory cortex sends a small number of fibers to the thalamus; but as shown in figure 7.1, most go directly to a special area called the *orbitofrontal cortex* (*ofc*) because it is situated just above the orbits of the eyes in the most anterior (prefrontal) part of the brain. It is shown in relation to the rest of the brain in figure 12.1.

The prefrontal cortex is regarded as the peak level of the primate and human brain, containing the circuits that subserve most of our highest human cognitive functions mentioned earlier. Astonishingly, the output from the olfactory cortex is aimed precisely at this highest level.

The sense of smell is therefore uniquely privileged in several ways:

1. It has a direct input to the prefrontal cortex.
2. It gets there through a short path involving only three neurons: olfactory receptor cells, mitral cells, and olfactory cortical pyramidal neurons.
3. Its area is situated in the heart of the part of the brain that makes us human.

The significance for neurogastronomes is clear: the volatile molecules released from everything we eat are so important that they are evaluated quickly at the highest level of the human brain.

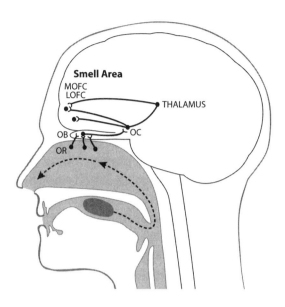

FIGURE 12.1 The human smell system
OR = olfactory receptor cells; OB = olfactory bulb; OC = olfactory cortex; MOFC = medial orbitofrontal cortex; LOFC = lateral orbitofrontal cortex.

Humans' Big Olfactory Brain

Maybe this direct pathway would not matter so much if the human sense of smell was not very important. Let us deal with the popular impression that humans have smaller olfactory brains than other mammals.

The bigger the brain, the more elaboration of sensory and motor representation can occur. This reaches its peak in the neocortex, the richest medium yet designed in animal life for elaboration of our sensory, motor, and internal worlds. It is usually assumed that the great overgrowth of the neocortex is related to the dominant role of vision in the lives of primates and humans. However, much of this great expansion of processing machinery is also available for the other senses as well, including smell. The main message of this book is that, despite the declining numbers of receptor genes, the brain processing mechanisms of the smell pathway, culminating in the neocortex, *bestow a richer world of smell and flavor on humans than on other animals.*

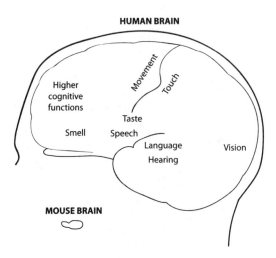

FIGURE 12.2 The human brain compared with the mouse brain for relative size
The diagram shows the locations of different brain systems related to different functions and behaviors.

This point can be made by comparing the mouse brain and the human brain. Because the two brains are usually illustrated at the same size, the olfactory bulb and orbitofrontal cortex of the mouse appear to be very large in relation to those of the human. However, this is where a picture, although it may be worth a thousand words, is worth the wrong words. If we show, as in figure 12.2, the two brains in their true sizes, the olfactory bulb of the human is actually nearly as big as the entire mouse brain, and the orbitofrontal region is huge by comparison. The conclusion is clear: the smell pathway has been maintained in size in the human, whereas the amount of brain to process the signals in that pathway has increased enormously. Much of neurogastronomy is about what this increased brain power can do to carry out enhanced processing of its smell and flavor input.

Unique Information Processing and Connectivity

What is there about this patch of neocortex that enables it to contribute a much enhanced human meaning to the smell image, in the form of a smell object arriving from the olfactory cortex?

There are two main possibilities. One is that its construction enables it to carry out enhanced processing of the smell input. The second is that its increased connections with other brain areas go far beyond what has been available to the olfactory cortex.

First, how is this "new" cortex constructed? It is thicker, because it represents a kind of doubling of the single layer of the simple olfactory cortex into two thick layers, containing more subvarieties of pyramidal cells and interneurons. In both layers the pyramidal cells have an organization similar to what we saw in the olfactory cortex: axon branches that feed back excitation on themselves and other pyramidal cells, and that also activate inhibitory interneurons to cause feedback and lateral inhibition. The result is that the individual basic operations, of boosting and recombining inputs and shaping them with inhibitory interneurons, appear to be similar in principle to those in the olfactory cortex, but greatly expanded in complexity. In addition, there is a new population of small *stellate* (star-shaped) cells, which serve (among other functions) as an internal relay for the sensory input coming either directly from the olfactory cortex or indirectly from the thalamus.

So we see again the principle that each stage of processing in the smell pathway is carried out by a new and more powerful type of processing mechanism.

Second, what about the connections? Recall that the connections of the olfactory cortex consisted of the input from the olfactory bulb and the output to the orbitofrontal cortex. The olfactory cortex is a region dedicated to processing the odor image from the olfactory bulb and outputting it to the orbitofrontal cortex. The olfactory cortex does have output connections to other parts of the brain besides the orbitofrontal cortex, but the only connections to the neocortex are the direct ones to the orbitofrontal cortex. In contrast, the orbitofrontal cortex has critical connections with many other parts of the neocortex.

Figure 12.3 shows how all the sensory pathways send projections from their neocortical areas to the orbitofrontal cortex. This means that the cells in the orbitofrontal cortex potentially are able to combine their olfactory inputs with inputs from all the other sensory systems that are stimulated by food in the mouth. They can also have interactions with other parts of the neocortex that are involved in other types of behavior. Chief among these are the *amygdala*, involved in emotion, and parts of the *prefrontal*

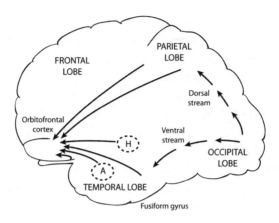

FIGURE 12.3 Many brain regions connect to the orbitofrontal cortex
H = hypothalamus; A = amygdala.

cortex, involved in flexible learning and in decisions about rewards. These are all capabilities that are necessary in judging flavors within the context of different senses in the mouth, different behavioral states such as whether we are hungry or satiated, and making decisions about whether a flavor is attractive or not, capabilities that are discussed further in later chapters.

Creating Smell

Given this smell pathway, how does it create the perception of a smell? Here we encounter a problem. Understanding the earlier stages of the olfactory pathway very effectively uses studies of laboratory animals such as rodents because there is a close resemblance between these stages in rodents and humans. However, in the case of the orbitofrontal cortex, these are two small areas in the rodent, whereas in the monkey and human they are relatively large, reflecting the much greater importance of neocortical processing of the olfactory input. We therefore have much less information about basic physiological properties of neocortical processing of smells than in other stages of the olfactory pathway. Our understanding at this level depends on studies of monkeys with cell recordings and of humans with functional imaging.

Work on this problem began in earnest in the 1970s with recordings from cells in the monkey by Sadayuki Takagi and his colleagues in Japan, showing that at successive stages in the olfactory pathway, from olfactory bulb to olfactory cortex to orbitofrontal cortex, the cell responses became more finely tuned to specific molecules. Then in the 1990s, a series of studies from the laboratory of Edmund Rolls at Oxford University showed how cells in the orbitofrontal cortex could respond to different combinations of tastes and smells. The advent of functional brain imaging in humans supported and extended these results. Out of these and subsequent studies, several important principles have emerged, which may be summarized as follows:

1. Some cells are tuned specifically to odors. These appear to represent the image of the smell object at this level and may constitute the neural basis of conscious smell perception.

2. Most cells do not respond to changes in odor intensity (in contrast to cells in the olfactory bulb and the olfactory cortex).

3. Some cells respond to both odors and taste stimuli. This can be called *sensory fusion*. It is not a property found in olfactory cortical cells and therefore appears to be the first step in creating the combined perception of flavor. (It has also been speculated that these combined responses could represent "sensory synesthesia," the unusual condition in which a person uses a term from one sense to describe a perception in another—as, for example, when a particular odor smells "blue.")

4. Some cells respond preferentially to pleasant smells, and others to unpleasant smells. The pleasant-responding cells tend to be localized in the medial part of the orbitofrontal cortex, and the unpleasant-responding ones tend to be in the lateral part of the orbitofrontal cortex, as shown with cell recordings and also with functional brain imaging. This separation is not seen at lower levels in the smell pathway.

The evidence that smells at this high level are arranged by the brain along a single axis, from pleasant to unpleasant, is summarized by Morten Kringelbach and Kent Berridge in their recent book *Pleasures of the Brain*. It is strongly supported by a theoretical approach of Noam Sobel and his colleagues in Israel, who believe that in this arrangement the complicated multidimensional nature of odor molecules is reduced to the unidimensional nature of odor objects. This attribute of smell is in turn one of the key attributes of the perception of flavor, a *quality axis* that obviously is not present in the senses themselves. In practical terms, when we consume our food and drink, we are not only detecting and discriminating among them, but also arranging them along our personal axis of pleasantness and unpleasantness. We will see that this applies to tasting wines, even by experts (chapter 24).

5. The qualities of pleasant and unpleasant are in turn reflective of their reward value, which we all know from our own flavor experience, and can be demonstrated in a monkey. We also know that we can change our preference from acceptance to rejection—for example, depending on whether we are hungry or full. Changing from acceptance to rejection of a smell can also be demonstrated in a monkey; it is known as *reversal learning.* Orbitofrontal cells are the first in the olfactory pathway to show this critical property of higher-level processing. It is an indication of how important smell is to flavor, because preferences for different smells can be learned and unlearned. It can be presumed that this, together with differences between smell receptors, is the basis for the development of preferences for smells by different cultures as well as the development of different individual preferences within a culture.

In summary, as Rolls states: "The . . . olfactory cortex represents the identity and intensity of odour in that activations there correlate with the subjective intensity of the odour, and the orbitofrontal and ACC [anterior cingulate cortex] represent the reward value of odour, in that activations there correlate with the subjective pleasantness of odour."

With smell perception at its core, we next consider the contributions of the other sensory systems in building the unified perception of flavor.

CHAPTER THIRTEEN

Taste and Flavor

The olfactory pathway can act alone when we smell the aromas of food or beverages by inhaling them by the orthonasal route. When smells arise from food in the mouth to make their contribution to flavor, they always act on the brain in concert with other systems. We need to keep reminding ourselves that in addition to the sense of smell being a dual system— orthonasal smell and retronasal smell—retronasal smell is never sensed by itself, but always together with virtually every other sense in the mouth. It is the basis for my claim that flavor is among the most complex of our perceptions created by our brains. As Frédéric Brochet and Denis Dubourdieu in Bordeaux note: "[T]he taste of a molecule, or of a blend of molecules, is constructed within the brain of a taster."

The most obvious sensory system contributing to "taste" is the taste system, the one that gets all the credit for the resulting flavor.

Taste Buds

As we all know, this system begins with stimulation by food and drink of the taste buds on the tongue and the back of the mouth. They are called *buds* because they consist of taste cells crammed together to make a kind of bud, or cartridge, with each sensory cell ending in fine hairs that carry receptors for different stimuli. The taste buds are located in different-size folds (papillae) of the tongue surface. If you look in a mirror with your tongue out, and especially if you paint it with a food dye, you can see small mushroomlike (fungiform) papillae in the middle of the tongue, medium folds (foliate papillae) on the sides at the back, and large round (circumvallate) papillae in the middle across the back.

The Molecular Basis of Taste

The taste buds contain cells that respond differentially to five kinds of stimuli. The traditional ones in Western culture have been salts, acids, sugars, and bitter compounds. Asian people traditionally identify a fifth type of stimulus, the amino acid glutamate, which they link to a meatlike perception called *umami*. This is now widely accepted as a basic taste, often termed *savory* or *meatlike*.

Each type of stimulus acts preferentially on a special type of receptor (box 13.1). Molecular biologists have used gene engineering to clone

BOX 13.1
Basic Taste Stimuli and Their Receptors

Salts act on a membrane channel that lets salt ions (sodium or potassium) flow through it.

Acids act on a membrane channel to modulate hydrogen ion flow through it.

Sugars act on sugar receptors (Taste Receptor Type 1, subtype 1 and 3: T1R2 and T1R3) that activate a second messenger system involving cyclic adenosine monophosphate (AMP).

Bitter compounds act on special bitter receptors that are also linked to a second messenger system called gustducin. The bitter receptor gene family (Taste Receptors Type 2: TAS2Rs) is numerous and diverse (more than 100 genes), reflecting their important role in detecting as many substances as possible that might be harmful to the organism.

Umami is due to the carboxylate anion of the amino acid L-glutamate (monosodium glutamate) found in broths and meats; it stimulates receptors called T1R1/T1R3 that are linked to second messengers.

and identify these different taste channels and receptors. Just as with the olfactory receptors, taste receptors are fundamental to molecular neurogastronomy.

Encoding Taste

The receptor cells in the taste bud have no axons to carry their responses to the brain, unlike the olfactory receptor cells. The cells interact with one another and with the endings of three types of nerve: the seventh, ninth, and tenth cranial nerves. The reason for mentioning them by their numbers is to highlight the fact that they are toward the back in the series of twelve cranial nerves beginning with the first nerve, the olfactory nerve.

The taste nerves enter the brain stem, the part of the brain that is directly continuous with the spinal cord and that is responsible for automatic functions such as control of heart rate, breathing, and other vital activities. There is thus a sharp contrast with the olfactory nerves that enter the front of the brain closest to the highest cognitive centers. It is as if the multisensory flavor sensation is designed to allow ingested foods to be analyzed at both the highest cognitive level as well as at the level of the most vital functions. It has all bets covered.

How the different kinds of taste are represented in the brain has stimulated two basic views. One, supported by psychophysicists who test subjects for their responses, is that each type of stimulus has its labeled line into the brain, leading to its distinct perception. This was called the *labeled line* theory. The other view, held by physiologists making recordings from the nerves, was that a nerve fiber tends to respond best to one type of taste stimulus, but that it also responds to two or three of the others to less of an extent. This is called an *across-fiber pattern* and indicates that there is a limited kind of combinatorial processing of the taste input. With gene engineering, Charles Zuker at Columbia University and other molecular biologists have been able to delete each type of receptor shown in box 13.1 and show that in most cases a given type of receptor is responsible for most (but not necessarily all) of its appropriate taste perception.

In the brain stem, the fibers connect to a group of cells called the *nucleus of the solitary tract*. From there the pathway in the primate goes to

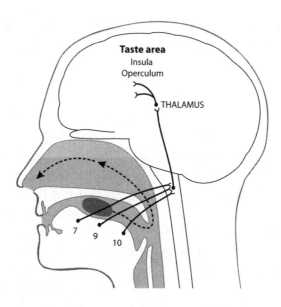

FIGURE 13.1 The human taste system
7 = facial nerve; 9 = glossopharyngeal nerve; 10 = vagus nerve.

the taste nucleus in the thalamus, where it is relayed to the taste area of the cerebral cortex. The primary cortical taste areas in the primate, including humans, are the anterior insula and nearby frontal operculum. *Insula* refers to the island of cortex inside the folds of the cerebral cortex; *operculum* refers to the folds that cover it. It is in these areas that conscious perception of taste is believed to occur (figure 13.1).

Taste Qualities

At the highest level of conscious perception, we classify the five basic taste sensations as salt, sour, sweet, bitter, and savory (umami)—the four *s*'s and a *b* (box 13.2). Note that these are not exactly the terms we used to describe the taste stimuli themselves. There is not, in fact, a direct mapping of those stimuli into corresponding sensations. For example, sweet has its major contribution from sugar receptors but also smaller contributions from the other types of receptors, and sour likewise has its major input from acid receptors but also contributions from

BOX 13.2
Basic Taste Perceptual Qualities and Significance

Saltiness is essential for maintaining salty body fluids derived from mammals' sea ancestry.
Sweetness is innate in all mammals, because sugars reliably signal high energy. Mother's milk and ripe fruits are sweet.
Sourness warns of food that may have gone bad.
Bitterness warns of toxic substances that should be rejected.
Savoriness is a meaty quality, signaling a high-energy food.

the other receptors. This reflects a limited combinatorial processing that takes place between the taste stimuli and the different taste receptor mechanisms, resulting in perceptual qualities that are multidimensional rather than one-dimensional.

Box 13.2 indicates only the simplest qualities associated with each type of taste stimulus, also called a "tastant." In fact, especially in human diets, any quality can be attractive, depending on the combination of qualities of a particular food or drink. Thus, although sweet is often considered the most attractive taste quality for humans, there are also among human cuisines preferences for salted meat, sour cream, bitter chocolate, and untold variations on these themes. Each of these tastes is accompanied by smell images that have become attractive through the learning process of growing up in a particular culture.

These basic tastes have powerful interactions with the other senses that make up flavor. Sweet and salt also strongly enhance food preferences, ranging from liking to craving. (For more on overeating associated with food that has an irresistible taste, see chapter 21.)

A Cortical Taste Map?

The ability to record the responses of nerve cells in cortical areas of the monkey has enabled investigators to confirm that cortical cells respond to tastants, presumably associated with the conscious responses. One might assume that because each tastant gives rise to its distinctive perception, there would be "labeled lines" in the taste pathway to spots in the taste cortex reserved for each quality. Thus far the evidence for such a map in the taste areas has been equivocal. So it is possible that at the cortical level the taste qualities are represented in a more associative manner similar to that in the olfactory cortex and the association cortical areas of other sensory systems.

Cortical Taste Responses

Each tastant has its threshold—its minimum concentration—for eliciting a response when given individually. However, as Justus Verhagen and Lina Engelen at the John B. Pierce Laboratory and Yale University explain in an extensive review in 2006 of these kinds of interactions, if two or more tastants are given together, there is an interesting effect: the threshold for the mixture is lowered, and the subject is able to detect the mixture when the individual tastants are weaker. This means that each weak tastant cannot be sensed by itself, but together they can be perceived (for example, salt and sour). This is called *intramodal enhancement*.

Cross-model enhancement occurs between sensory systems. This is particularly interesting for taste and smell interactions. A weak smell and weak taste can be subthreshold by themselves, but together they can be sensed, although usually only if they are "congruent"—that is, if they complement each other. Congruency may be innate, or it may be learned. Noncongruent smell and taste stimuli do not have this mutually enhancing effect.

The effect of one stimulus enhancing another is believed to be of survival value for the animal. For humans, it has the interesting implication that the more sensory elements there are in our foods, the more it enhances their flavor. Our traditional cuisines consist of multiple different sensory stimuli that, one may hypothesize, usually are congruent.

In contrast, modern "molecular gastronomy" often explicitly experiments with combinations of senses that are not congruent. In these cases it may well be that two or more different senses in a food are not congruent and therefore do not enhance each other. Conversely, this incongruent effect may be precisely the aim of the mixture: to give an entirely new and different flavor perception.

Taste and Retronasal Smell Together

Taste stimuli usually occur together with retronasal smells. In fact, smell and taste together are often regarded as forming the main basis of flavor. If it is important to be able to detect this co-stimulation, it should not be surprising that orbitofrontal neurons are found that respond to both taste and smell stimuli. The two sensations are so closely related that in some cases smells take on the qualities of taste, as when one says that something smells sweet. This is called *sensory fusion*, regarded by some as an example of synesthesia, defined in chapter 12 as interpreting the quality of one type of sensory stimulus in terms of another sense: an odor smells sweet; a sweet taste is blue.

The cells in these primary taste areas have been shown to be sensitive not only to taste but also to other senses, such as texture and temperature. Some cells respond only to taste, whereas others respond to the other modalities. We see here the neural signs of how the cortex builds the multimodal nature of flavor.

Apart from the effect of taste and retronasal smell together on responses in the primary cortical areas, a striking finding is that together they also enlarge the amount of cerebral cortex that is activated. This was shown in a classic experiment by Dana Small and her colleagues in 2005. They first stimulated subjects separately with a retronasal smell and with a tastant, which activated the two primary cortical receiving areas, as shown by functional imaging. They then delivered the two stimuli simultaneously, as would happen during ingestion of a food. This activated not only the primary areas but also previously unstimulated secondary surrounding areas of the cortex. Presumably these added areas were recruited to analyze the more complex mixed stimuli. The cortical responses combine taste and smell to create the unified quality of flavor.

Taste and Emotion

We know that tastes are strong elicitors of emotion. The most marked facial expressions of pleasure or disgust come from tasting foods that please or repulse. The facial expressions reflect what is called the *hedonic* (emotional) quality of the food.

Are these hedonic qualities hardwired or are they learned? Jacob Steiner, a pediatrician in Israel, set out in 1974 to test whether the hedonic properties of the different tastants are present at birth—in other words, whether they are hardwired or have to be learned by experience. He arranged to test newborn infants in a hospital in the first hour or two after birth before they had exposure to any taste, and to film their facial expressions.

The results were dramatic. When testing with a cotton swab dipped in a salt solution, the infants showed little facial response, just looking around and wondering what a curious world they had been born into. When tested with an acid, they showed a definite aversive puckering of the lips. A bitter substance produced a face of disgust and distress with an open mouth trying to eject the substance. Finally, it all turned out wonderfully, because sugar elicited a fleeting but unmistakable smile.

The implications of this experiment are quite interesting. They show that the basic emotional (hedonic) qualities of the tastants are hardwired at birth. The infant does not have to learn that sugar is good, salt is fine, acid is not so good, and bitter is to be rejected. This makes adaptive sense for the survival of the organism. But even more interesting is that the basic emotions of happiness and unhappiness are present at birth. We do not have to learn how to express these fundamental emotions. This also makes adaptive sense for the bonding of infant and parent. Expressions of affection and disapproval are unambiguous in forming the new relation, and they continue in their basic form throughout childhood and adult life.

These results in the human were confirmed by Ralph Norgren and his colleagues at Penn State University in experiments on infant rat pups, which show the same types of responses to the same types of stimulants. Furthermore, in the rats it was possible to test more directly the important question of whether the hedonic responses are mediated entirely at the brain stem level or might include some higher brain processing, by carrying out experiments removing just the brain cortex. The basic responses were not affected, showing that the hedonic responses are hardwired only at the brain stem level. As we noted, this is where the taste fibers enter the central nervous system.

In summary, the taste system builds hardwired emotional expressions for its attractive and repellent stimuli. This means that when we experience a flavor we really like, our happiness is expressed through facial expressions laid down at birth and co-opted by smell. This provides a foundation for further consideration of the emotions of flavor in chapter 19.

Supertasters

We assume that everyone has the same sensations of taste we do, but this is not true. Amazingly, Jean Anthelme Brillat-Savarin already knew this:

> I have already stated that the sense of taste resides mainly in the papillae of the tongue. Now the study of anatomy teaches us that all tongues are not equally endowed with these taste buds, so that some may possess even three times as many of them as others. This circumstance explains why, of two diners seated at the same table, one is delightfully affected by it, while the other seems almost to force himself to eat; the latter has a tongue but thinly paved with papillae, which proves that the empire of taste may also have its blind and its deaf subjects.

Brillat-Savarin is describing what we now call *individual variation.* He was remarkably prescient. The most studied variation in taste is the ability to taste the bitterness of a particular bitter compound called *propylthiouracil,* or PROP for short. Linda Bartoshuk, formerly at Yale and now at the University of Florida, coined the term *supertasters,* characterized as individuals who taste PROP as extremely bitter and those who have the highest density of fungiform papillae on the tongue tip.

Bartoshuk, her longtime collaborator Valerie Duffy at the University of Connecticut, and other colleagues have built on the groundwork laid by Roland Fischer in the 1960s to explain variability in food preferences, alcohol consumption, and body weight, based on their data from PROP tasting, fungiform papillae density, and other taste markers, such as the bitterness of quinine. Steven Wooding and Dennis Drayna at the National Institutes of Health in 2004 discovered the gene that is responsible for 70 percent of the variation in the bitterness of PROP. Since then, more of the taste receptor genes have been paired with the ability to taste bitterness of specific bitter compounds, including coffee and grapefruit juice.

According to Duffy, she and others are studying a variety of tastes and taste genes to identify individual variation in taste and link this variation to diet and health outcomes. For example, nontasters to PROP or those carrying receptors with lower response to bitter compounds appear to like fat, sweet foods, and alcoholic beverages; are heavier; and are at greater risk of alcohol overconsumption. By contrast, supertasters report greater disliking for vegetables and tend to be thinner. PROP sensitivity is associated with a higher taste sensitivity in general, and with a larger number of taste buds on the tongue. You can inspect your tongue in the mirror and see if you are a supertaster. If you are, be assured that this is a normal variation in the population. But it is something to keep in mind when you sit down to your next dinner: Are you experiencing more or less of the taste than your dinner companions? In sum, to paraphrase Brillat-Savarin: Tell me your bitter receptor genes, and I will tell you who you are.

Bait Shyness

An important aspect of taste is that it is the critical line of defense against the ingestion of stuff that would be bad for us. Stuff that is poisonous is usually bitter, for which the bitter taste receptors are tuned. But what about a pleasant-tasting food that turns out to have harbored an infective agent that makes an animal sick or was used as bait to lure an animal into a trap? In a famous study, John Garcia and Robert Koelling showed in 1966 that an animal that has been made sick from a food just once will avoid that food ever after, even though the sickness occurs many hours later. It is called *conditioned taste aversion*; in field studies of animals it is called *bait shyness*. Because it requires only one episode of sickness, it is also called *one-trial learning*.

Bait shyness is so much more powerful than classical learning, which involves many paired associations between stimuli, that at first few psychologists believed it. One of them who reviewed Garcia and Koelling's first article was quoted as commenting: "These findings are no more likely than finding bird s—— in a cuckoo clock!" But we are all familiar with this kind of learning. When we suspect a food has made us sick, we lose our taste for it and search through our memories for the cause of it. It has also been called the *Béarnaise sauce phenomenon* after causing sickness in a well-known psychologist. In subtle forms it may lurk behind food phobias that develop during childhood and in some cases last a lifetime.

CHAPTER FOURTEEN

Mouth-Sense and Flavor

In addition to having a smell and a taste, every kind of food and drink has physical properties that give it a feel in the mouth. This feel is the result of so many senses that it is difficult to summarize in one name, so we call it variously *mouth-sense*, *mouth-feel*, or *food-texture*. This sensation is something we prejudge when we take the food to consume it. We compare this expectation against the actual sensation in the mouth as part of judging the quality of the food. Mouth-sense is mediated by the somatosensory (*somato = body*) system, which encompasses a range of sensory submodalities that include touch, pressure, temperature, and pain. These somatosensory fibers add many important sensory qualities to the taste in our mouths and the smell in our noses, combining with them in the unified perception of flavor.

The Receptors in the Mouth

There are several kinds of somatosensory receptors in the skin of the lips and cheek and inside the mouth on its membranes and on the tongue (box 14.1). The receptors are similar to those in other parts of the body, except that they are present in very high density. Obviously the mouth cares a great deal about the feel, the texture, the temperature, and the potential noxious quality of what it takes in.

BOX 14.1
Somatosensory (Mouth-sense) Receptors

Temperature-sensitive terminals come from very small slowly conducting nerve fibers. They have molecular receptors called Transient Receptor Potential (TRP: pronounced "trip") channels that can sense cooling or warming of the skin or membrane. They sense hot food and coffee, ice cream and cold drinks.

Pain-sensitive terminals come from small fibers. The painful stimuli may be mechanical, chemical, or temperature. TRP channels have been implicated in some of these responses. Depending on how the terminals are activated, they can give sticking pain, burning pain, aching pain, and almost any other kind of pain you have experienced or can imagine. They are what give many their delight or unhappiness on consuming hot peppers (capsaicin).

Touch-sensitive terminals come from medium-size faster nerve fibers. They can sense small depressions, roughness, and stretching of the skin and membrane, such as while we are chewing. TRP channels appear to be involved here, too.

Pressure-sensitive terminals (Pacinian corpuscles) are extremely sensitive to mechanical vibration. They respond quickly but transiently to pressure such as vibration from rapid tongue and jaw movements.

The Receptors in the Nose

Not to be forgotten are the somatosensory fibers in the nose. In analogy with the mouth, we can say that they are responsible for "nose-sense" or "nose-feel." These are thin nerve fibers with receptors in their terminals in the membrane lining the nasal cavity. They are activated by certain kinds of volatile chemicals, such as the vapor of ammonia; by physical properties such as those of carbonated drinks; by a rapid rush of air through the nasal cavity when sniffing; and by moderate-to-high concentrations of many kinds of smell molecules. Psychologists studying the psychophysical properties of different smell molecules have to control carefully for whether a smell perception is due partly to these somato-sensory receptors in addition to the smell receptors. There is evidence that the smell receptor cells can also respond to mechanical stimulation by the air stream, making smell itself a multimodal sense. So except for very quiet breathing, somatosensation is always present when we smell.

The Many Textures of Food

The array of sensory receptors in our lips, mouths, and tongues is matched by the array of physical and chemical properties of the foods and drinks we put in our mouths. This interaction gives rise to the large number of sensory qualities that we call the mouth-feel or mouth-sense: the touch, feel, and texture of the food and liquids in our mouths. Here we note just a few of them.

As box 14.2 indicates, the somatosensory system is critical for testing the state of the food. Humans place a high value on the many mouth-feel properties we look for in different foods. Free endings in the tongue and membranes of the cheeks and palate, and even in the teeth, signal crispness, smoothness, chunkiness, stiffness, softness, springiness, rate of fragmentation as we chew, and readiness to be swallowed—to name just

BOX 14.2
Qualities of Mouth-Sense: Touch, Feel, and Texture

Smooth	Soft	Pain (burning)
Creamy	Slippery	Pain (sticking)
Viscous (thick)	Rough	Pain (aching)
Crunchy	Gritty	Ease of fragmentation
Crisp	Astringent	Sticky
Springy	Hot	Dry
Chunky	Cold	Crumbly
Stiff	Lukewarm	

a few qualities. The acceptability of French fries, for example, depends on how crisp they are as well as on their taste and smell. Few enjoy breakfast if the cereal flakes have become limp in the humidity because someone neglected to close the package tightly (one of my pet gripes).

Pain commonly signals something to be avoided. It is used, together with the touch receptors, to reject sharp objects like fish bones that could be life threatening if swallowed. Anyone who has eaten whole fish knows how thoroughly the fish meat is turned over and over in the mouth so that any tiny bone can be detected by a sharp touch or pricking pain and rejected.

Pain is a special case for flavor. Most people reject a food that causes pain in the mouth, particularly burning pain. However, many people learn to tolerate and prize the pain wrought by the capsaicin molecules in chili peppers as an essential part of the flavor of the foods they are in. This is an example of how the hardwired aversive behavioral responses at birth can be overcome by learning.

The Touch Pathway to the Cortex

The receptors for all these qualities are on the tips of nerve fibers that belong to the fifth cranial nerve, called the *trigeminal nerve*. From the nose and mouth, the impulses travel to the trigeminal ganglion and into the trigeminal nucleus in the brain stem (figure 14.1). Like taste fibers, their entry is at the level of control of vital functions, as one might expect for an input carrying signals of pain that require an immediate reaction.

The output cells send their axons to the somatosensory nucleus in the thalamus, from where the axons are relayed to the somatosensory area of the cerebral cortex. The fact that the sensory endings in the mouth and tongue are densely packed implies a correspondingly large amount of cortex to analyze these responses, and that is exactly what is found. The neurosurgeon Wilder Penfield in Montreal showed this in the 1950s. He carried out the first operations on humans for relief of epilepsy, which involved opening the brain case to expose the entire cortex surface. This enabled him to locate the sensory and motor areas so that he could avoid interfering with them. He was able to map out the body

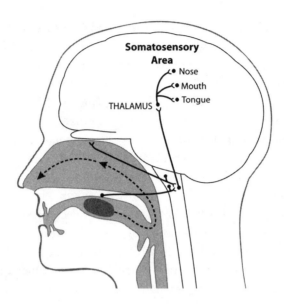

FIGURE 14.1 The human somatosensory system

surface on a strip of cortex and show that there is a regular sequence from the toes to the head, and that it varies in size depending on how sensitive the body parts are to touch. The largest areas were the fingers, lips, and tongue. The fingers bring food to the lips, which together with the tongue and other receptors in the mouth sense it while it is being chewed and swallowed; these actions represent a large proportion of the somatosensory part of the brain, all devoted to the touch dimension of flavor. An increase in activity of these areas in obese individuals will be discussed in chapter 21.

It appears that conscious localization of the touch sensations to the nose and mouth occurs at this neocortical level. Just as we can relate individual types of smell molecules to their spatial patterns in the olfactory bulb, so can we relate specific types of touch sensations to their receptors in the mouth. However, the sensations that contribute to the unified sensation of flavor must involve further connections between those areas and the cortical areas for taste and smell.

Interactions of Touch with Taste and Smell

We have indicated that flavor is a multisensory sensation. It is now becoming apparent that it involves much more than simply summing up the different senses that contribute to it. They interact in many ways with each other, beginning with the receptors and continuing all the way up to the cerebral cortex. Some of these effects and interactions that contribute to the perception of flavor have been reviewed by Justus Verhagen and Lina Engelen in 2006 and are summarized in box 14.3.

These interactions are only a few of the many that have been studied in the laboratory and that are, of course, known in traditional cuisines. It hardly needs to be pointed out that it is precisely these interactions that are the focus of so much interest in the manipulations of different food compounds that underlie the drive to create new kinds of cuisines, including the sometimes noncongruent ingredients that are explored in molecular gastronomy. It seems inevitable that increased knowledge of the brain mechanisms involved will enhance these explorations.

BOX 14.3
Interactions of Touch with Taste and Smell

Effects Between Submodalities of Touch
Temperature on touch: Differences in temperature affect the perception of the texture of a substance in the mouth. These perceptual differences may be due to peripheral or central effects.
Temperature on irritation: The hot burning quality of pain produced by capsaicin is reduced by cooling and increased by warming.
Irritation on temperature: Irritants, such as capsaicin, increase the perceived intensity of warmth and reduce the perception of coolness.

(*continued*)

Effects Between Touch and Taste

Taste on touch: Sweet makes substances feel more viscous (thicker), whereas sour makes them feel less viscous.

Touch on taste: Stimulating one part of the tongue containing taste buds and then another part without taste buds makes the subject think that the stimulus is working in both parts. This is known as *tactile capture*. It reflects the fact that every aspect of flavor is referred to where the stimulus is located.

Temperature on taste: Sensitivity to tastants is highest at normal mouth temperature (between 71.6 and 98.6°F [22 and 37°C]).

Taste on irritation: Pain due to capsaicin reduces sensitivity to the different tastes (sweet, sour, salt, bitter, and umami).

Temperature and viscosity on taste: Increasing heat releases more taste molecules from a substance in the mouth, which—combined with lower viscosity from the heat—facilitates stimulation of the taste buds. This is why hot food tastes stronger, especially cooked meats. These are physicochemical effects in the mouth that are relayed up the pathways to the cortex to give stronger perceptions of flavor.

Effects Between Touch and Smell

Touch on smell: Increasing the viscosity of a gel in the mouth reduces the perception of retronasal smell. This could be due to a peripheral interaction (less release of smell molecules) or central interaction (convergence of sensory inputs in the cortex).

Temperature on smell: Increasing heat releases more volatile compounds and hence stimulates the olfactory receptors more strongly by retronasal smell. This is one reason you want your coffee hot.

Smell on irritation: Orthonasal odor stimulation inhibits orthonasal irritating stimulation.

Irritation on smell: When delivered together, the irritant suppresses the odor. Stimulation with capsaicin inhibits the perception of flavors.

CHAPTER FIFTEEN

Seeing and Flavor

We all know that if we are hungry, the sight of a tasty food makes our mouths water. Technically speaking, the sight stimulates our autonomic nerves to activate our salivary glands to secrete saliva into our mouths to prepare to ingest and digest the food. We usually don't consider that the sight might also have an influence on the flavor of the food, but much common experience and many experiments have shown this to be the case. So, although sight is not a property of the food once it is in our mouths, it is part of the multisensory sensation of flavor that the food produces. This is important, both for understanding the flavors of the food we eat and for understanding how advertisers manipulate the visual impact of food and drink to influence how much we like them.

The Visual Pathway

Humans are highly visual animals, and this is clearly evident in how dominant the visual pathway is in our brains.

Vision starts with our eyes, where light falling on the retina activates photoreceptor cells to form images of the visual world. Two thin layers of neurons within the retina take this image and begin to process it, to extract different properties of the visual images. These include registering how light and dark the images are, enhancing the contrast between light and dark (chapter 6), responding to different wavelengths, beginning to create colors from the different wavelengths (chapter 9), registering the sizes of objects, and responding to direction of movement of an image across the retina. In addition, the two eyes work together to register depth of focus of an object.

The main lesson from the retina is thus that instead of just one type of stimulus—one modality such as the identity of molecules as we saw with the sense of smell—the visual pathway from the start has to process a half-dozen or more types of stimuli. Neuroscientists are showing how it does this by its incredibly precise neuronal microcircuits, which overlap in their functions to build a united image.

All this information is projected through the optic nerve that emerges from the back of the eye through a crossover point (the optic chiasm) where the nerves from the inner half of the retina connect to the other side to create the unified image of the two eyes. These fibers end in the thalamus, in the so-called lateral geniculate nucleus. From there, fibers arise that project to the primary visual area of the neocortex at the back of the brain. Higher processing of the visual image begins in earnest there and proceeds through several stages involving surrounding "association" cortical areas. Cell recordings show neurons tuned to different properties: direction of movement, contrast between black and white, contrast between different colors, and depth perception. There is strong evidence that these are the neural substrates underlying the psychological perceptions of these properties.

To compare this pathway with that for smell, we should recall from chapter 10 that for smell, only two synapses (in the olfactory bulb and the olfactory cortex) lie between the receptor cells and the primary neocortical area in the orbitofrontal cortex, whereas we can trace visual processing through as many as four or five synaptic relays (two in the retina, one in the thalamus, and one to four in the neocortex) to produce the psychological properties of conscious vision. The reason for this has been explained to me by a computational neuroscientist. Visual processing is expensive; it involves processing many different submodalities, each of which requires much computation. For example, think of depth perception, which is essential for life on land or in the trees. It requires computing in three dimensions the locations of objects, their direction of movement, their changes in size with movement, their speed of movement, their shape, their color, and so forth. These are all things you take for granted as you drive your car on the freeway at 70 miles (around 110 kilometers) an hour, while your visual brain is doing the work.

For our interest in comparing vision with smell and flavor, an important point is that all this processing, through these several steps at the neocortical level, takes place entirely with respect to the visual image. In comparison, we've seen that much processing of the smell image takes place at the level of the olfactory bulb and olfactory cortex. When the smell image arrives at the neocortical area in the orbitofrontal cortex, it isn't further processed as an exclusively smell image. Instead it immediately interacts with other sensory systems, such as taste and touch, and with other systems involved in emotion and memory.

Visual processing does eventually involve interconnections with the prefrontal cortex; in fact, it is believed that it is through these connections that consciousness occurs, as will be discussed in chapter 25. These are part of the converging connections from other sensory systems shown in figure 12.1. For now, our interest is in whether any of these visual properties interact with those of smell, taste, or mouth-sense to affect our perception of flavor. Of course, the shape and size of a flavorful object are among the properties we use in identifying a food object and judging its attractiveness. However, the most important property is color.

Color and Flavor

Color especially has an influence on the sense of smell. In an early experiment by Tryg Engen in 1972 at Brown University, subjects were tested with odorless solutions that were either colored or uncolored; the subjects tended to rate the colored solutions as having an odor. This was interpreted to indicate that when we see a colored solution we expect it to have a smell. The test was performed using only orthonasal sniffing, not retronasal smell.

This line of investigation was pursued by Deborah Zellner and her colleagues in 2005 at Montclair State University. They presented smell stimuli in solutions that were either colored or colorless. Using orthonasal sniffing, the subjects judged the colored ones to smell more intense than the uncolored ones. This result fits with Engen's study and with common experience: a more colorful solution makes the solution smell more intense. Because colored solutions like fruit juices usually have distinctive smells, seeing them stimulates the expectancy of the smell by itself. Some psychologists regard this as a kind of Pavlovian conditioning. Just as Pavlov's dogs expected flavorful meat when it was paired with a bell, so we expect a juice to be flavorful when it has the appropriate color.

A similar result was expected when this same experiment was carried out on retronasal smell, but there was a surprise. The subjects were told to swallow the solutions, colored or uncolored, and rate the intensity of the retronasal smell. Unexpectedly, the retronasal smell of a colored solution was experienced as less intense than that of an uncolored solution. In some experiments, the subjects could experience the smell orthonasally as they took the cup of solution to their mouths to drink it, so it was speculated that the subjects expected the solution to be more intense than it actually was. The explanation is that the color-induced heightened intensity of the orthonasal sensation made the retronasal sensation seem suppressed. This is a good example of how complicated the relation between a behavioral property and its brain mechanisms may be.

The inverse of this experiment is sensing unpleasant smells. Paul Rozin and his colleagues in 1995 found that a food with objectionable smells was more objectionable when taken in the mouth and sensed retronasally than when tested orthonasally. This makes sense; the mouth is the guardian of what we take into our bodies, and an objectionable smell and flavor is riskier than a pleasant one.

These experiments reemphasize that orthonasal and retronasal smell are different submodalities of the same modality, different senses within the same sense, each with its properties reflecting the different routes of stimulation, different associations with other senses, and different perceptions.

Wine Color and Flavor

Dramatic examples of the effect of color on flavor have come from studies of wine tasting. Many experimenters and wine critics have shown that the characterization of a wine usually begins with its color. As Frédéric Brochet and Denis Dubourdieu of the Faculty of Oenology at the University of Bordeaux comment:

> [D]escriptive terms are always connected to the color of the wine. . . . This can be seen as the brain's necessity to retrieve a strong correlation to the world it perceives and describes in language. Wine flavor, which is a complex mixture, is then described using words characteristic of objects having the same color. Thus color is the only common categorization among subjects.

Perhaps the most dramatic effect of color on smell and flavor discrimination was shown in a classic test of red and white wines carried out by Gil Morrot, Frédéric Brochet, and Denis Dubourdieu in 2001. They demonstrated that white wines falsely colored red were mistaken by wine tasters for red wines. Because this experiment has become well known, we will consider it in detail.

For this experiment, Bordeaux AOC wines of vintage 1996 were used: a red made from cabernet-sauvignon and merlot grapes, and a white made from semillion and sauvignon grapes. A first test was to compare the flavors of the real red and white, and a second test was to compare the flavor of the same white wine, colored red with a grape anthocyanin, with the white. Separate tests with blindfolded subjects showed that this colored white was in fact indistinguishable in taste from the normal white. So the experiment was to see if the subjects would be influenced by the color of the wine in discriminating a red from a white wine, when the red was a red-colored white.

The tests were performed on 54 undergraduates enrolled in the Faculty of Oenology in the University of Bordeaux, so they were not yet experts but they did have a definite academic interest in wines. They tasted the wines in glasses under artificial white lighting in individual booths. They were allowed to compare the wines ad libitum.

A problem was how to deal with the expected individual differences in the ways people describe wines, which, according to the experimenters, "are too great to establish a representative odor space for a whole group." To deal with this, the experimenters supplied a list of terms for describing the flavor sensations. The subjects could use these terms, called *descriptors*, or they could use their own. The descriptors were mostly floral or fruity terms, such as *pear*, *pineapple*, *honey*, *woody*, and *cassis*.

In a first round of tests, the subjects had to indicate which of the real red or white wines fit each of the descriptors best. These tests were carried out on the real red and white. The subjects then came back a week later for a second round. For this, each subject was given his or her descriptors from the first round, in alphabetical order, and was required to use them for describing the wines again. The subjects assumed that they were testing the same wines again to see how consistent they were, but for this round, without knowing it, their comparison was between the false red-colored white and the same white. Again, they "were asked to indicate for each descriptor of their list which of the two wines most intensely presented the character of this descriptor."

How could the subjects have misjudged the colored whites to be reds? The same whites they had previously characterized as white? The interpretation of this experiment is subtle, because it involves several factors.

First is the question of attention. In the second round, the subjects were focused on distinguishing between wines they assumed were the same red and white they had tested in the first round. Their attention was therefore on using the same descriptors for what they believed was the same purpose.

It is well known that attention to one aspect of a problem can make one oblivious to other aspects that may be altered. A striking example is the video of two people at a table exchanging objects between them; the test is to tell at the end who has which object. During this time, out of view as the camera shifts from one person to the other, one person puts on a different-color shirt. Yet at the end when asked, the observers are oblivious of this change. Similarly, a video of two people in a crowded room asks the observers to relate what goes on between them. Meanwhile a person in a gorilla suit wanders through the background; again, this does not register with the observers. Attention, therefore, plays a powerful role in determining what we observe.

Second is the question of language. As already discussed, Dubourdieu and his group are experts in the lexical analysis of wine tasting—that is, the analysis of how words are used for describing the flavors of wines. In fact, the color experiment was as much a test of the language used for the descriptors as it was for the effect of color. In previous work, they showed that wine experts use language in special ways to characterize wines:

> All wine descriptive language is in fact organized around wine types which we call prototypes. If this is in fact correct, what a wine taster does in front of a wine is not an analysis of its separate sensory properties but a comparison of all the cognitive associations he or she has from the wine (color, initial aroma, and taste) with the impressions he or she has already experienced when tasting other wines.

In the final section of their paper, Brochet and Dubourdieu pose the question: "What does this reveal about brain function?"

First, single-cell recordings in the orbitofrontal cortex show neurons responding to both taste and smell, as well as convergence of submodalties of touch. Brochet and Dubourdieu comment: "This could explain the inability of wine tasters to separate olfactory and taste description because these data are mixed physiologically at an early perceptive stage." We have explained the basis for this sensory fusion in the chapters on responses to smell, taste, and touch of neurons in the orbitofrontal cortex.

Second, they cite the work of Dana Small and her colleagues in an article in 1997 that showed that responses to flavor occur mainly in the right hemisphere, which is implicated in "ideogram perception and dimensional representation," whereas more analytical processes occur mainly in the left hemisphere. The concept of ideograms is equivalent to that of smell images that we have been using to characterize the representation of smell stimuli in the brain.

And third, the olfactory areas of the orbitofrontal cortex have connections, among other areas, with the hypothalamus. This gives an ultimate basis for every judgment of smell (and flavor) in what we've seen experts call "hedonics"—that is, things that give one pleasure:

> Basically, animals use the sense of smell to know whether they can eat a food. This is probably what humans do as well. The main cognitive concern regarding flavors is whether they are good or not. This concern is so strong that even experts cannot ignore it and it is what drives the organization of their descriptive language. In this way experts are not so different from novices.

What is described here is essentially the axis of pleasant to unpleasant we discussed in chapter 11 that is a major way that the orbitofrontal cortex characterizes flavors. It is reassuring to know that when it comes to deciding what we like, we are all on a level playing field with the experts!

Our excursion into color has thus led us into a deeper insight into the importance of language in flavor, which is discussed further in chapter 25.

CHAPTER SIXTEEN

Hearing and Flavor

The last sensory system we will discuss that contributes to flavor is hearing. Although it seems unlikely, it has its own role to play.

The Auditory Pathway

The auditory pathway starts with the ear, composed of the outer ear and the ear canal leading to the ear drum (tympanic membrane). Inside is the middle ear, with the three little bones that transfer the vibration of the eardrum to the round window of the inner ear, where the cochlea contains the hair cells that respond to the vibrations set up by the sounds.

The hair cells pass their responses on to the endings of ganglion cells that connect to the cochlear nucleus in the brain stem. Here the fibers make contact with a number of different types of neuron, each of which begins to extract specific attributes of the auditory stimuli, such as pitch, loudness, or duration; in other words, as in the case of smell, the neurons begin to create our human experience of sound. This information then enters a complex network of brain stem centers that process the signals and send them to a higher brain stem center, the inferior colliculus, which in turn sends them to the auditory thalamus, and from there on to the primary auditory neocortical receiving area. Thus, in contrast to olfaction and vision, hearing, like taste and touch, starts as a brain stem sense. However, at the cortical level, it quickly establishes itself in humans as a higher cognitive sense through its role in speech and language.

How Food Sounds

From this brief review it appears that the auditory system is designed to receive sound signals from the environment. That is how we think of it in our daily lives. However, from the point of view of flavor, the system is relevant for the sounds it picks up as we consume our food and drinks.

From an evolutionary point of view, we can assume that the sounds of food being bitten and chewed gave significant information to our ancestors about the toughness of a vegetable, the ripeness of a fruit, or the squishiness of a hunk of flesh.

In our everyday lives, we don't usually think of the sound of food as a part of its flavor, but in fact it is. The "snap, crackle, and pop" of a breakfast cereal can be as important a selling point as its taste; in fact, the qualities are inseparable in making up the flavor. The crunching sound of deep-fried French-fried potatoes or chicken nuggets is an integral part of the chewing experience. And it is not just the sound of the food itself. The sound of our jaw working our mandibular joint is to our eating experience as familiar as the sound of a backhoe to its driver in digging in the ground. Although it is not essential to our eating experience, it is part of what we expect of it. And if it's not there, we notice it.

Although research on the interaction of hearing with the other senses of flavor is limited, some interesting facts are known. For example, crispness is a quality desirable in many foods. Some studies have claimed that crispness is the food texture of which people are most aware. This would make sense if it was critical for judging the ripeness of fruit for our human ancestors. One might assume that it is a quality related to touch and jaw action. However, an early study suggested that crispness of a food is judged primarily from the way it sounds as the teeth crush the brittle foodstuff, such as a potato chip. Presumably a crisp sound makes the chip taste better. This seems as though the sensation might be interpreted at the neocortical level.

Other studies have discriminated between crispness and "crackliness," and between crispness and crunchiness. *Crisp* has been applied to the chewing sounds of flat breads, with high frequency sounds above

5 kHz. *Crunchy* has been applied to the sound of chewing raw carrots, which produces lower pitches of 1 to 2 kHz. *Crackly* has been used to characterize dry biscuits, which produce lower frequencies, sensed more through bone conduction in the mandible and maxilla. In general, the clearer and louder the sound, the more we like it.

Several of the mouth-feel stimuli have their characteristic sounds in the mouth; such as highly viscous foods compared with thin soups.

The Sound of Wine

Liquids have their sounds as they are swished in the mouth or swallowed. Wine tasters use every cue to tell them about the quality of a wine, as this anecdote I heard in France illustrates:

> The chef Paul Bocuse defines thusly the ideal wine: it satisfies perfectly all five senses: vision, by its color; smell, by its bouquet; touch, by its freshness; taste, by its flavor; and hearing, by its *glou-glou*.

Bocuse, apparently in jest, includes the sense of hearing as being essential to the ideal wine by using the term *glou-glou* to describe the sound of the wine being swallowed. Although it seems like modern slang, *glou-glou* in fact is a classical reference in French literature, as I learned when I told this anecdote to my friend Jacques Guicharnaud. A courtly man with a lively sense of humor, Guicharnaud was a much-loved teacher of French literature at Yale University for many years. After I told him Bocuse's anecdote, he pounced and said "Oh, *glou-glou* is a quote from Molière." And indeed it is. In *The Doctor in Spite of Himself* (*Le médecin malgré lui*), Sganarelle comes out drunk, holding high a bottle and singing a little song:

> How sweet from you
> My bottle true;
> How sweet from you
> Your little glouglou.
>   *Act 1, scene 5*

The *glou-glou* sound of the wine disappearing down our throat is due to the muscle activity we use in moving our tongue about and swallowing, reminding us that flavor depends on our motor systems as well as our senses.

CHAPTER SEVENTEEN

The Muscles of Flavor

When we talk about gastronomy, people usually mean how the food stimulates our senses to give rise to particular flavors. However, this requires moving the food within our mouths in a coordinated manner. We take this part of eating for granted, but closer inspection shows that it is incredibly complicated. I'd like to convince you it is one of our supreme motor acts. The next time you take a bite, think of what is happening in your mouth. There is a complex sequence of movements involving the muscles of the lips, jaw, and tongue and those involved in swallowing; these movements have to be coordinated with breathing, and all of them must be coordinated while we are focusing our senses on enjoying the flavors that are produced.

We are not the first to be interested in understanding the mouth as a motor organ. On this subject, Jean Anthelme Brillat-Savarin was particularly eloquent:

> I have discovered at least three movements [of the tongue] which are unknown to animals, and which I describe as movements of SPICA-TION, ROTATION, AND VERRITION (from the Latin *verro*, I sweep). The first takes place when the tip of the tongue protrudes between the lips which squeeze it; the second, when it rolls around in the space between the cheeks and the palate; the third, when it catches, by curving itself now up and now down, the particles of food which have stuck in the semicircular moat between the lips and the gums.

The reader may try these tongue motions in order to verify the savant's claims.

This theme has been taken up by two modern-day disciples of Brillat-Savarin. A vivid description of the extraordinary range of movements involved in manipulating food has been given by a well-known chef, Jean-Marie Amat, and a leading neuroscientist and gourmet, Jean-Didier Vincent, in their book *L'art de parler la bouche pleine* (*The Art of Talking with Your Mouth Full*). In one passage of the book, they write of the culinary adventures of a band of five good friends who gather every month or so to eat and talk—often, as the title indicates, attempting both at the same time. Most of the talk is about food and the human species that consumes it. At one point, the following spirited exchange takes place:

> The human mouth, what a marvel, exclaimed the Professor, there's no animal that matches it!
>
> I don't detest a cow's facial appearance, replied the Artist. The way a cow chews its cud defines its way of life, its serenity in dealing with its world.
>
> Of course! responded the Professor. And what about the big cats? They're matchless in tearing meat apart. And dogs? Without equal in cleaning bones. But no beast anywhere combines as many talents as the human in being aware of its nourishment: chewing, crushing, manipulating, as well as "spication," rotating, "verrition"; and at the core, the tongue, that admirable muscle, veritable prayer rug of the palate, which by its delicate texture and its neighboring membranes expresses the sublimity of the operations for which they are destined. It is therefore not surprising that the orifice which enables an animal to nourish itself should become in humans at the same time the organ of language, the hallmark of the species and of the meal, which is the foundation of human society.
>
> Be that as it may, you're talking with your mouth full, interjected the Merchant. Couldn't you restrain your oratorical ardor?
>
> The splutter is the salt of eloquence, observed the Tobacconist.

Here we find emphasized, in more elegant words and narrative, several points we have been making about the physiology of eating.

First is the amazing range of ways that food is manipulated in the mouth, which reflects the much greater interest that a human takes in the sensory nature of the meal. The varieties of ways the food is moved about within the mouth of course enhance its sensory qualities, especially the retronasal smells, but also cause multiple exposures to the taste buds and to the multiple receptors for mouth-sense in the inner walls of the oral cavity.

Second, the Professor draws attention to the fact that the tongue is critical both to the sensory experience of the meal and to the production of language. It seems, in fact, paradoxical that the tongue should be so closely related to producing both smells and words, because it is difficult to find words for a smell. The explanation appears to lie in the concept of the smell image, as discussed in chapter 24.

Finally, the Professor makes the claim that the meal is at the heart of human society. Just as for Grethe and me, the gathering of a core family or an extended family or clan at mealtime to share the catch of the day was surely a key activity in forming human societies and cultures, and therefore must have played a key role in human evolution. This idea is discussed further in chapter 26.

Building on this eloquent testimonial to the movements within the mouth that produce the flavors of our food, let us take a look at what modern science is telling us about the motor systems that move our mouths. This evidence comes first from study of the central motor pathways in the brain, and second from using modern video techniques to observe the exact sequence of movements while consuming foods and beverages. There are also important clues gleaned from studies by anthropologists of our human ancestors. The following draws on *The Evolution of the Human Head*, a book by Daniel Lieberman that describes in exquisite detail the movements of the human mouth in eating.

The Flavor Motor System

The descending motor system that runs our muscles and glands begins at the highest level in the motor strip of the cerebral cortex. Just as there is a sensory homunculus representing the body surface, with large areas devoted to the lips and tongue, there is a corresponding motor homunculus with a similar enlargement of the lips and tongue. It represents the larger numbers of cortical microcircuits devoted to receiving the sensory inputs and controlling the fine movements of the lips and tongue when we eat and drink.

From this high level, the output neurons—large pyramidal neurons controlled by the same kinds of feedback excitatory and inhibitory lateral connections found in all cortical areas—send their fibers all the way down through the interior of the brain and into the brain stem to terminate directly or indirectly on the large motor neurons that control the muscles of the lips and tongue (figure 17.1).

This motor system is an essential part of the flavor system. The lips and tongue are constantly moving as we eat and drink. We have noted that the sensation of flavor seems to be coming from the mouth, even though much of the flavor is due to retronasal smell. This is called *mouth capture*. In addition to the sensory signals, mouth capture is believed to be due also to the high level of activity in the tongue and lips areas of the

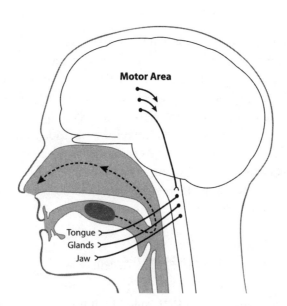

FIGURE 17.1 The human motor system to the mouth area

motor cortex. It all adds to the illusion we live under when we enjoy the flavor of our food and give the mouth all the credit.

How Chewing Produces Flavor

Let us discuss four types of movement within the mouth that are activated and controlled by the descending motor system involved in producing flavor: chewing, swishing, swallowing, and breathing. What does evolution tell us about their roles in producing human flavors?

The evolutionary record leaves few traces of the things that really interest us about how we became human, but they do give us fascinating glimpses relevant to flavor. This is because the only parts of our human ancestors that have survived are the bones, and of these the bones of the head are among the hardest, and of those the teeth are the hardest of all. The patterns of wear on the teeth tell anthropologists much about what people ate; this is usually used to infer what the diet was. Our interest is in how chewing produces flavor.

Mammalian teeth come in several varieties. In front are the incisors, sharp and bladelike, and the canines, pointed like daggers. We use both to grip, cut, and tear meat and vegetables into morsels. This action requires considerable force, exerted by the jaw muscles. The size and placement of the muscles contribute greatly to the form of the human face.

The food morsels are then moved by the tongue to the premolars and molars, where chewing begins. The tongue directs the traffic between the food mass in the mouth and the moving teeth, feeding small bits between the teeth for grinding into smaller bits and finally into a mush (now called a *bolus*). This grinding (now called *mastication*) crushes the food cells, releasing the volatile molecules of the morsels of meat or fruit or vegetables. Like all mammals, we chew mostly on one side in order to concentrate more force. This requires the tongue to keep the chewed mass on the one side. Most foods require many chewing repetitions. The comedian Bob Newhart used to insist that his mother told him to use as many chews as he had teeth in his mouth. Another rule of thumb is that the most flavor is obtained (and the bolus is in its most digestible form) when the individual types of food in the bolus are no longer distinguishable by the touch receptors on the tongue.

Reducing the time and effort for chewing was one of the key steps in human evolution. Our ape cousins spend half the day chewing tough meat or vegetable matter. As discussed in chapter 4, there is considerable evidence that cooking was one of the most important steps in human evolution, because it tenderized not only meats but also fruits and vegetables, increasing their nutritive value while reducing the chewing time.

How Swishing Produces Flavor

Consuming fluids is absolutely necessary for all mammalian life, including primates and humans. Fluids may enter the mouth in various ways: for most animals by lapping up from a pool or a dish, and for humans from various types of containers or through straws. The lips enable the transfer from the source to the mouth. The fluid may spend very little time in the mouth, for example, when one is thirsty. Or the fluid may spend quite a while in the mouth, as in the drinking of wine. In this case, the tongue comes into play in a different way, swishing the fluid about slowly in order to present it to the taste buds on the tongue in as many different ways as possible, while allowing prolonged sensing and evaluation of the retronasal smells that emanate from the fluid. It is probably not an overstatement to suggest that the expertise of a wine connoisseur is highly dependent on the particular tongue movements that have been learned in presenting the wine in an optimal manner for sensory evaluation.

How Swallowing Works

As the bolus reaches its choice consistency, the tongue moves it toward the back of the mouth, where there is a final evaluation through the touch receptors of whether it is ready to be swallowed. Then an exquisitely coordinated movement takes place. The soft palate is stiffened and raised to seal off the part of the nasopharynx that mediates retronasal smell. The tongue moves the bolus into the pharynx at the back of the mouth and then rises to force the bolus down into the pharynx. At the same time, muscles relax in the lower pharynx to allow the bolus to pass into the esophagus, and other muscles contract to close the vocal folds and raise the epiglottis so that the bolus cannot go the wrong way down the trachea. Finally, peristaltic movements move the bolus down the esophagus to pass into the stomach. Food physiologists are studying these events in ever greater detail. For example, Andrew Taylor and his colleagues at the University of Norttingham in England have analyzed flavor release during chewing and swallowing of different amounts of food with differing fat content. They find relatively more flavor release from larger amounts of low-fat food. How much we stuff into our mouths is thus another variable in the complex perception of flavor.

How Breathing Produces Flavor

Breathing produces flavor in two ways: during mastication and after swallowing. During mastication, the bolus is in the mouth, and as long as it stays there we can continue breathing, sampling the volatiles emitted from the bolus with each expiration, together with the stimulation of the other senses (taste and touch). Videofluorography is being used to document the exquisite coordination required for breathing during mastication, pharyngeal bolus aggregation, and swallowing. You can document this for yourself by observing each step the next time you are eating your dinner. During the steps of swallowing, the bolus leaves the tongue, so there is no taste, and the airway is blocked off, so there is no retrograde smell. During the swallow we hold our breath to prevent food from going down our windpipe. However, as soon as the swallow is over, an exhalation takes place. It is usually completely automatic, even if we have not inhaled before. We may continue to sample the flavor with continued breathing. In that case, what is being sensed is the thin layer of food remaining on the pharyngeal walls.

Similar steps apply to fluids, as reported by Taylor and his colleagues in a recent study of aroma release during drinking wine. We breathe while swishing a sip of wine in our mouths, the flavor sensation coming from the volatiles when we exhale.

In summary, these many coordinated muscle movements mean that flavor is an "active" sense. Other senses may be stimulated in a passive way, merely by exposing us to a sight, a sound, or a smell. However, flavor always must result from movements involved in taking the food into our mouths, masticating it, and swallowing it. All the movements originate in motor systems distributed widely in the brain, further emphasizing how much of the human brain is involved in the seemingly simple acts of enjoying what we are eating and drinking.

CHAPTER EIGHTEEN

Putting It Together
The Human Brain Flavor System

We have covered the main sensory and motor systems of the brain that are involved in generating the sensation of "taste," which we now recognize is really "flavor." We share these systems with other animals, but there are some big differences that characterize humans. We have fewer olfactory receptors, but we are much more adapted to retronasal smell, and we have much bigger brains, with many more brain areas and connections between them. And we have language. Our overall brain system for flavor therefore varies in quantitative terms from that of other animals, and qualitatively it has new capacities to create and elaborate the sensation of flavor. To recognize this, I will call it the *human brain flavor system* and claim that it is uniquely human.

That humans have a unique brain system for flavor was already predicted by Jean Anthelme Brillat-Savarin. In his book's early section "Pleasures Caused by Taste" (*taste* standing here for *flavor*, as we have often noted), he waxes eloquent:

> [T]aste as Nature has endowed us with it is still that one of our senses which gives us the greatest joy:
>
> (1) Because the pleasure of eating is the only one which, indulged in moderately, is not followed by regret;

(2) Because it is common to all periods in history, all ages of man, and all social conditions;

(3) Because it recurs of necessity at least once every day, and can be repeated without inconvenience two or three times in that space of hours;

(4) Because it can mingle with all the other pleasures, and even console us for their absence;

(5) Because its sensations are at once more lasting than others and more subject to our will;

(6) Because, finally, in eating we experience a certain special and indefinable well-being, which arises from our instinctive realization that by the very act we perform we are repairing our bodily losses and prolonging our lives.

And in the following section, "The Supremacy of Man," he writes:

[O]f all the creatures who walk, swim, climb, or fly, man is the one whose sense of taste is the most perfect.

The tongue of an animal is comparable in its sensitivity to his intelligence.

The idea that only humans experience pleasures that are perfect as opposed to those of other animals was a conceit of nineteenth-century thought, reflecting the conviction that humans are at the pinnacle of the animal world in whatever we do. Now in our post-Darwinian era we would characterize each animal in terms of its shared general properties through common or parallel lineages, and by those properties that are unique to it in finding its evolutionary niche. From this perspective, we would recast Brillat-Savarin's claims by saying that the human brain has specific capabilities that make the appreciation of flavor of unique importance in humans. Brillat-Savarin in fact hinted at this in writing: "[S]ince taste must not be weighed except by the nature of the sensation which it arouses in the center of life, an impression received by an animal cannot be compared with one felt by a man."

The human brain flavor system can be thought of as containing two stages. The first stage is the sensory systems that feed into it, which we have covered so far. These transform the individual sensory representations into the combined sense of flavor. The second stage is the action systems that draw on the full capacity of the human brain systems that generate and control our behavior. The systems we have considered in the previous chapters are collected in the diagram of figure 18.1. This gives a better impression than the individual diagrams of the extensive amount of the human brain devoted to flavor. It will be useful to summarize this sensory stage as a preparation for explaining the action system in the following chapters.

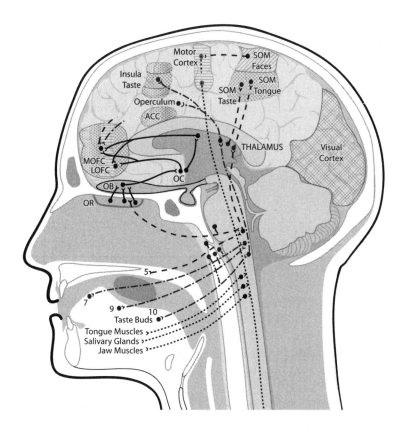

FIGURE 18.1 The human brain flavor system summarized from previous figures
Abbreviations are the same as those in the previous illustrations of the different systems. ACC = anterior cingulate cortex; SOM = somatosensory cortex.
(Adapted from G. M. Shepherd, Smell images and the flavour system in the human brain, *Nature* 444 [2006]: 316–321)

Flavor Perception System

As we have seen, the human brain flavor system begins with the five senses receiving their stimuli in their receptors and converting them into neural representations:

• *Smell*. Among these inputs, smell is unique in going directly to the olfactory cortex in the forebrain limbic system, where it forms distributed memories of the smell stimuli represented as odor objects. Within the limbic system the smell objects therefore have direct access to brain systems for memory and emotion. The olfactory cortex further projects to the orbitofrontal cortex at the front of the brain, where it connects to the highest centers concerned with the uniquely human capacities for judgment and planning.

• *Taste*. The pathways of the different tastes plunge into the brain stem, where they immediately have access to hardwired expressions of the emotional qualities of the taste stimuli. They proceed further to their cortical areas, where they interact with all the other sensory representations at the core of flavor.

• *Mouth-sense*. The different types of touch that food and liquid in the mouth activate are sent through the touch pathways in the brain stem to the thalamus and their cortical receiving and association areas. The mouth and tongue have an enormous representation in the cortex, which accounts for how dominant our perception of the food in our mouths is as well as the illusion that its smell is coming from its taste.

• *Sight*. The sight of our food and drink before we consume it activates the visual pathway that passes through the thalamus to the visual areas at the back of the brain. It has a highly significant influence on how we judge its flavor, as advertisers well know in enticing us to eat what they produce.

• *Sound*. Finally, the sound of our food as we eat it is an integral part of the flavor experience.

One of the questions I find most interesting is how the multiple sensory pathways indicated in figure 18.1 interact to create the flavor we sense.

Multisensory integration occurs when the cell responses in a region to two or more stimuli at the same time are more than the addition of the individual responses. This is a property called *supra-additivity*. It can occur in cells in several regions when congruent (complementary) stimuli are delivered simultaneously.

The behavioral response is a perception of flavor from the combined senses, such as taste and smell stimuli when they are congruent. Congruency is a property that is learned, presumably involving the mechanisms in the olfactory cortex that projects to the orbitofrontal cortex and through it to the other regions such as the insula and anterior cingulate gyrus. Maria Veldhuizen, Kristen Rudenga, and Dana Small at the John B. Pierce Laboratory and Yale University suggest that the fact that the "supra-additive responses in these regions are experience-dependent strongly supports the possibility that these areas are key nodes of the distributed representations of the flavor object."

The concept of a "distributed representation of the flavor object" is useful. This distributed activity constitutes, in the brain, an internal "image" of the flavor of the food or drink produced by stimulating the sensory receptors in the nose and mouth. Just as the sensors are distributed in the body and gather different types of information about the food, so the brain's internal representation of the stimulating object arises from activity in different brain regions, each with its particular combination of input and output connections.

These considerations lead to the idea of an "internal brain image" of the flavor object. Ideas about such internal images have long been dominated by vision. A visual mental image is distinct from the sensory visual image that is projected onto the retina and transmitted through the thalamus to the visual cortex. The two images meet somewhere in the primary and secondary association visual areas, a zone referred to as a *buffer*. This is where the bottom-up image that we see with our eyes meets the top-down image that we imagine.

"Mental images" are usually considered within a given modality, such as visual images (as when one imagines a visual object that one examines from different angles). Other examples are "smell images" as mental images (as when one imagines a smell), "auditory images" (as when one imagines a song), and even "motor images" (as when one imagines a willed movement). The idea of "odor imagery" has received support from studies of human brain scans by M. Bensafi and Noam Sobel and their colleagues. These examples of internal mental states depicting the imagined object seem intuitively reasonable, and in fact can be related to actual sites in the brain. In the case of vision, there is evidence that the buffer between external and internal images is located in the visual cortex.

This "pictorial" view represents the most widely accepted concept of mental images. Stephen Kosslyn, at Harvard University and now at the University of California, Berkeley, has been a leading advocate of this majority view. An alternative minority view is that our internal "image" is not in a literal pictorial form but rather in a "propositional" form that is more appropriate for representing the action that the image engenders.

In this discussion we are extending the idea of a mental image from one that is limited to within a brain area devoted to one modality, such as the visual cortex, to one being distributed among several different brain areas representing multiple modalities. A consensus is emerging that simultaneous activation by a food of a common set of regions, including the *orbitofrontal cortex, anterior insula and overlying operculum, frontal operculum, and anterior cingulate gyrus*, constitutes the distributed representation in our minds of a flavor object. We can express this by saying that the "simultaneous supra-additive learned activity in these regions constitutes a flavor image." This perceptual image makes up the neural representation of a remembered flavor.

Flavor Action System

This, then, is the first part of the flavor system, the part that produces the perception of flavor. What we do with that perception—how it engages our behavioral responses—is the second part. I call this the *flavor action system*. The true extent of the human flavor system is revealed in the network of regions and connections that reflects the power of the greatly expanded human brain to give full meaning to the flavor image and elicit characteristically human behaviors. The left side of figure 18.2 indicates the sensory systems we have already discussed. The right side of the figure indicates the main parts of the brain that mediate these behaviors. As with the representation of the sensory systems in figure 18.1, this diagram

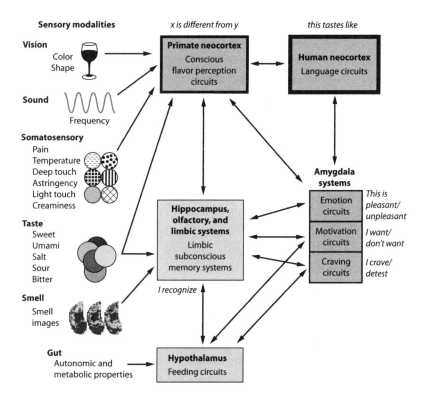

FIGURE 18.2 The human brain flavor system shown as a flow diagram
(Adapted from G. M. Shepherd, Smell images and the flavour system in the human brain, *Nature* 444 [2006]: 316–321)

shows the great extent of the human brain devoted to the action systems of flavor. The chapters that follow will consider each of these systems.

• *Emotion.* All mammals have brain systems for emotion. In humans, these are highly developed in size and in their extent of interconnectedness. As a result, our emotions expand from those related directly to the flavor sensation to engage regions involved in all aspects of our mental life.

• *Memory.* All mammals also have brain systems for memory, and they are very good; the memory of an elephant is legendary. In humans there are many regions of the brain for the memories laid down by the flavor system to engage. This is illustrated by the iconic tale of Proust and the remembrance brought about by the flavor of a madeleine biscuit.

• *Decisions.* One of the hallmarks of our bigger brains is our power to decide among complex alternatives that affect the present and the future. Nowhere are the decisions more important than in choosing the balance between flavor and nutrition in what we eat.

• *Plasticity.* What our brains do affects what our brains become. This dependence of our brains on activity reflects the plasticity that is built into our neural cells and connections. We need to know how our flavor preferences have long-term effects in changing the flavor system itself.

• *Language.* Humans are unique in having language, and this ability greatly affects how we sense and judge flavor.

• *Consciousness.* The neural basis of consciousness is a hot topic in neuroscience and philosophy. Most of the debate is about what we consciously "see," but the neural basis of conscious smell and flavor are challenging problems yet to be solved.

Let us begin with this question: Why do we like the flavors we like? What explains the close coupling between the flavor images and the pleasure we get from these images? The explanation lies in the connections that are made with the emotional centers of the brain.

PART IV

Why It Matters

CHAPTER NINETEEN

Flavor and Emotions

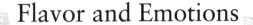

In moving from the sensory input to the action output within the human brain flavor system, it is natural to begin with emotions. The word *emotions* is derived from "to move." Just as movements of the mouth and tongue make flavor an active sense, so is it also an active sense in that we must be motivated to acquire the food and liquid we put in our mouths. As indicated in the previous chapter, these systems have a close relation to the part of the brain called the amygdala, and their activity can be seen as beginning with the motivation for *wanting* a food or liquid, which may be learned as a *liking* for it and then become a *craving* for it.

What kind of brain activity represents our motivation to desire a flavor, our emotion that makes us prefer it and want to obtain it? And if we become too highly motivated, too desiring, how does this brain activity pass over to craving it? These questions have stimulated much research that it is hoped will help us understand normal liking for food and abnormal craving for it.

Images of Desire

In the human, investigators have used functional brain imaging to answer these questions. Among the first to do so were Marcia Pelchat and her collaborators at the Monell Chemical Senses Institute in Philadelphia in 2004. It will be useful to describe this pioneering work in detail

---

BOX 19.1
Brain Regions Activated by Pleasant Food Smells

Orbitofrontal cortex	Cingulate	Parahippocampal gyrus
Insula	Amygdala	Anterior fusiform gyrus
		Striatum

---

as an example of how perceptions interact with emotions and how this interaction is studied in human subjects.

The authors begin by pointing out that craving a favorite food is experienced by most people, particularly young people, and may play a significant role in excessive snacking, eating disorders, and obesity. There is no strong evidence that cravings are for nutritional types of food, but there is evidence that a dull, boring diet stimulates strong cravings for more flavorful foods.

The authors noted that studies up to then had shown that food flavors activate certain brain areas. These areas were largely consistent with our discussion in the previous chapter and are summarized in box 19.1. These previous studies used hunger to stimulate the desire for the food. Hunger affects the whole body in a way that may obscure pure cravings for a desired food. The authors therefore decided instead to use a monotonous diet as a baseline for judging the activation of brain regions caused by craving.

At the time no one had done this kind of study. One hypothesis was that the areas activated by craved foods might be the same ones activated by pleasant foods. However, there are many foods we like without craving them. In addition, the authors were well aware that craving is also a characteristic of drug addiction. A great many people had been interested in the brain mechanisms underlying drug addiction and had worked out experimental procedures for bringing out drug cravings in patients, which could be studied using brain imaging. These activities had produced evidence that cravings for drugs, including alcohol, activated specific areas in the brain, which are summarized in box 19.2.

The overlap of the first four items in each table was intriguing. So the authors decided to focus on possible similarities in brain circuits when

BOX 19.2
Brain Regions Activated by Drugs of Abuse

Orbitofrontal cortex	Dorsolateral prefrontal cortex
Insula	Hippocampus
Anterior cingulate	Caudate
Amygdala	

subjects were stimulated by craved foods and by craved drugs: "We hypothesize that individuals receiving a monotonous diet will have greater food cravings and related activation in these candidate regions than individuals maintained on their normal diet." The hypothesis was bold in suggesting that a strong desire for particular foods we like—an exaggeration of otherwise normal feeding habits—might activate the same brain circuits as an abnormal addiction to substances of abuse, including social drugs such as tobacco and alcohol as well as illegal drugs such as cocaine. It was the kind of hypothesis that drives science, because it enabled the investigators to structure their study in an effective manner that not only was well focused on the subject at hand, but also allowed direct comparisons with the other studies.

The subjects were healthy college volunteers, some of whom were maintained for a couple of days on a normal diet and some on a monotonous diet. The monotonous diet consisted of a vanilla-flavored drink containing 240 kcalories plus protein and vitamins. The subjects consumed, on average, nine 8-ounce (226-gram) cans of the stuff a day, for a total of around 2,200 calories per day.

At the end of the second day, the subjects were tested in the brain scanner. They were first tested at rest. Then names of two foods they had selected that they really liked were presented to them visually, and they were asked to imagine the smell, taste, and texture of their favorite dish of the food while the brain scanner recorded their activity. Presenting the names instead of the pictures of the food avoided showing nonoptimal versions of the food to different subjects; with the names, the subjects could themselves imagine their favorite version. The functional images of what is called the BOLD signal were obtained in a strong

BOX 19.3
Brain Regions Activated by Craving a Food

Left hippocampus
Left insula
Right caudate nucleus

magnet (rated at 4 Tesla) that enabled activity in small brain regions to be seen.

The subjects all reported that they could easily imagine their favorite foods. The subjects on the monotonous diet in addition reported that they felt a craving for their liked food while imagining it, whereas craving was only reported by a portion of the subjects on a normal diet. No one reported a craving for the monotonous diet.

The brain scans were unequivocal. The subjects on the monotonous diet showed activation of specific brain regions while imagining their favorite foods, whereas those on a normal diet did not. The activated regions are shown in box 19.3. This result was significant, because all these regions fall within the group activated by drug cravings. Only the insula is shared with the pleasant food smells.

The authors thus consider that the results support their hypothesis of "a common circuitry for desire for natural and pathological rewards."

The hippocampus, the insula, and the caudate nucleus are three areas that also merit further discussion in the interest of neurogastronomy.

The *hippocampus* has been shown to be involved in cocaine addiction, possibly by reinvoking memory mechanisms that drive this behavior. It may similarly provide the memory traces that drive food craving. In psychological terms, the memory functions as the reinforcer for the learned craving. When the subject sees the drug, the incoming visual image is reinforced by the memory trace in the brain. The sites where they meet would be equivalent to the buffer of a mental image, as discussed in the previous chapter.

We have already met the insula as a site of convergence of taste and smell inputs; it has also been shown to be involved in taste memories and in emotional behavior. Like the hippocampus, it may contain memories that act as reinforcers.

The *caudate* nucleus is a part of the system known as the *striatum* and plays multiple critical roles in sensorimotor coordination in the brain. It contains a high concentration of dopamine, the neurotransmitter released by fibers from the so-called substantia nigra (a region in the brain stem that appeared black to the early histologists who discovered it). This system is vulnerable to a range of disorders. Degeneration of the dopamine-containing cells in the substantia nigra and the terminals of their fibers in the striatum causes Parkinson's disease, which is associated with tremor of the hands, slowing of movement, and a range of other disorders, including (surprisingly) a decline in the sense of smell. Disorders of dopamine are also believed to be involved in schizophrenia.

Among the normal functions of the striatum are its involvement in the formation of motor habits, which is of interest; food cravings can be regarded as habits that are hard to break. Dopamine is also involved in the reward system of the orbitofrontal cortex, which we will discuss in chapters 21 and 22. With regard to drug cravings, it has been suggested by Ingmar Franken, Jan Booij, and Wim van den Brink in the Netherlands that "in conditioned subjects dopamine has a role [in] the earlier, motivational phase, i.e., before the use of the drug and before the experience of pleasure per se. This motivational phase can be labeled as the desire phase of drug use: [in other words] craving." This comment emphasizes the complex relations between the brain systems that produce the brain states we call motivation, leading to desire and craving, as well as to pleasure.

Finally, by subtracting the brain foci due to monotonous food cues from foci of liked food cues, the investigators were able to identify more regions that were activated when the subjects only thought about liking a food as well as craving it, as summarized in box 19.4.

Among these are the following that should be of interest for understanding the brain mechanisms for creating not only the perception of a food or flavor but the motivational and emotional states that make us like it.

The *left fusiform gyrus* on the bottom of the temporal lobe is interesting, because many studies have shown that activity in this area reflects

BOX 19.4
Brain Regions Activated by Liking a Food

Left fusiform gyrus	Bilateral cingulate gyrus
Left parahippocampal gyrus	Left caudate nucleus
Right amygdala	Right putamen

our perception of an emotion we are expressing. It works together with the nearby amygdala and parahippocampal gyrus in the following way. When hungry subjects view pictures of food, functional brain scans show activity specifically in these areas, as reported by Marsel Mesulam and his colleagues at the Northwestern Medical School in Chicago in 2001. The authors interpret their results as follows: "These results support the hypothesis that the amygdala and associated inferotemporal regions [the lower parts of the temporal lobe] are involved in the integration of subjective interoceptive [inside the body] states with relevant sensory cues processed along the ventral visual stream [the parts of the visual system that are involved in identifying objects]."

In plain language, this means that when we are hungry and view a picture of a food that we like, the picture sets up activity in the visual pathway that reaches cells in these regions to produce our personal internal "food image" of that food. This "interoceptive" image produces an emotion of liking those foods, as well as a motivation to acquire and consume them. It is an example of a mental image that is distributed among different regions and different modalities, a "multiregional multimodal image" that represents emotional and motivational states rather than the perception of what is pictured. The implication for the neurogastronome is that when you sit down to enjoy your meal, the hungrier you are the more active are your internal "emotional flavor images" of the food flavors you are perceiving. Pelchat and her colleagues call these *images of desire*. They are to the flavor action system what the flavor images are to the flavor sensory systems.

The *amygdala* is usually more involved than indicated here. As we have seen, it is sensitive to the intensity of taste stimuli, among other things. It also has an essential role in a wide range of emotional behavior in all mammals.

The *cingulate gyrus* is also of special interest to neurogastronomy. It is a ring of cortex just above the fibers of the corpus callosum that interconnect the two hemispheres. Several brain-scanning studies have shown that this is the brain region most consistently activated by stimuli that have a strong emotional quality and that figure in both food liking and food and drug craving. We have seen that it is part of several of the frontal lobe systems involved in those states.

Finally, the authors note that activity in the *orbitofrontal cortex* does not show up in their results. This may be due to the difficulty of imaging the BOLD signal in this part of the cortex, which is so close to the underlying bony braincase. But they note it may also be due to the fact that orbitofrontal activity may be involved in thinking about food in general, whether it is liked or disliked. This only reminds us that a scientific study combines biology, experimental methods, and interpretation, leaving unexplained results that require further study.

The overlap between cravings for food and drugs of addiction has become a major theme in modern studies, as is discussed in detail in chapter 22.

Chocolate: From Craving to Disgust!

A second perspective on brain mechanisms in food craving comes from a study by Dana Small and her colleagues in 2001 on craving for chocolate. They recruited nine subjects who described themselves as "chocolate lovers" and were high on a scale of "chocoholics." Four and a half hours after breakfast, they began to be tested in a brain scanner with a square of chocolate. The diabolic strategy behind the study was not only to examine the brain areas activated by the chocolate at the start, but to continue the experiment with successive squares of chocolate until the subjects had been fed to satiety—until they could not stand to eat another square of chocolate.

The subjects varied, some of them quitting when they reached 16 squares (about a half bar of chocolate), whereas others lasted up to 74 squares

BOX 19.5
Brain Regions Activated by Chocolate

Like
Subcallosal region
Orbitofrontal cortex
Insula and operculum
Striatum
Midbrain

Dislike
Orbitofrontal cortex
Parahippocampal gyrus
Prefrontal regions

(about two and a half bars). There was a five-minute rest period between each square. As the study progressed, from hunger to satiety, the subjects were asked with each square how it rated on a scale of "delicious, I really want another piece" to "awful, eating more would make me sick."

The brain scans thus provided evidence regarding the brain areas that were active when subjects were hungry and highly motivated to eat a craved-for food and those that were active when the subjects forced themselves to eat the same food when it tasted "awful." The results are summarized in box 19.5. At the beginning, the active brain areas were the subcallosal region [under the corpus callosum connecting the two hemispheres], the orbitofrontal cortex, the insula and operculum, the striatum, and the midbrain. We can call this the *flavor image of chocolate when it is desired*. At the end, the active areas were the orbitofrontal cortex, the parahippocampal gyrus, and the prefrontal regions. This can be called the *flavor image of chocolate when it has been eaten to satiety*. Neurogastronomes may use this as an indication of how the activity within our brains shifts with the consumption of even our favorite foods.

A subtle aspect of this research strategy was that what changed was the *reward value* of the chocolate. There is much current interest in brain science in the reward value of a stimulus, because it signals what an animal holds important, is motivated to work for, and will make decisions about acquiring or not acquiring in relation to alternatives. Wolfram Schultz in Switzerland introduced this notion with his studies of how the dopamine system in the brain is activated when a monkey is making such decisions about reward value. The chocolate study was an example of making decisions about reward value in relation to hunger or satiety. As discussed in chapter 12, reward value is one of the key functions of the orbitofrontal cortex, the neocortical end station of the olfactory pathway. This will be discussed further when we consider how the brain makes decisions about flavor (chapter 22).

CHAPTER TWENTY

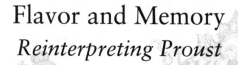

Flavor and Memory
Reinterpreting Proust

For many people, the most important parts of smell and flavor are the memories they evoke and the emotions associated with them. To illustrate the impact these have, we can do no better than start with Marcel Proust. An understanding of the brain mechanisms involved throws new light on his classic story:

> But when from a long-distant past nothing subsists, after the people are dead, after the things are broken and scattered, taste and smell alone, more fragile but more enduring, more unsubstantial, more persistent, more faithful, remain poised a long time, like souls, remembering, waiting, hoping, amid the ruins of all the rest; and bear unflinchingly, in the tiny and almost impalpable drop of their essence, the vast structure of recollection.

The quotation comes from *Swann's Way,* the first volume of his great novel *Remembrance of Things Past.* In it, Proust describes how from the aroma of a tea-soaked biscuit called a madeleine a powerful memory of his childhood flooded back to him. This has become a cliché for the memory that springs suddenly and purely into mind after a long period of forgetting. However, several years ago a young English professor of my acquaintance carried out a close textual analysis that revealed that the memory was not sudden, but was the result of considerable effort as described by the author himself. That made me start to wonder what was happening during this time in Proust's brain, between the initial sensing of the taste of the tea-soaked madeleine biscuit and the memory that was finally called forth.

An understanding of the neural pathways in the sense of smell and the perception of flavor starts to provide an answer. Several years ago, Kirsten Shepherd-Barr and I combined the textual analysis and the neural pathways to reinterpret this famous passage, and I update that reinterpretation here. If you have followed the argument in preceding chapters, you will recognize the principles as they unfold; if you have plunged in here, you will get the whole story as it applies to this episode.

Activating Proust's Brain

We start with the knowledge that the taste of a madeleine must be mostly due to its smell. The stimulus for Proust's taste experience was therefore primarily the odors emanating from the mixture of pastry crumbs soaked in *tilleul*, the aromatic lime-scented infusion made from linden blossoms. What, then, might these smells have been?

It is sometimes argued that Proust dithered over exactly what kind of biscuit it was that stimulated his reverie, but it really does not matter. A traditionally made madeleine, in addition to possessing odor molecules that arise from the butter and eggs, would include several types of "aroma essences." Flavors of foods are enhanced by heating and dissolving in water, which increase the vapor pressure so that volatile molecules are released into the air or within the mouth. Thus, as children learn, a humble pastry gives off its aromas with greater effect when its crumbs are dissolved in hot liquid. The aromas in a madeleine would include vanilla and several types of related odor molecules in the lemon, such as citral and limonene, which belong to the terpenes, a family of essential oils secreted by plants. As described in chapter 4, they are highly volatile, consisting of 5 carbon atom units linked together in various shapes and with various reactive functional groups such as esters, alcohols, and acids. The other source of olfactory stimulation in Proust's concoction was the *tilleul*, which contributed its own scent.

How do these molecules give rise to a smell perception? As we have seen, receptor molecules in the fine hairlike cilia lying in the mucus are stimulated by the smell molecules as they are inhaled into the nose (the orthonasal route), and also by the smell molecules released from within the mouth that rise into the nasal cavity from the back of the mouth (the retronasal route). It is by this latter route, after the narrator has taken the brew into his mouth, that the smell molecules are released and carried by the warm and humid air of his nasopharynx to his olfactory sensory cells.

The smell molecules, absorbed into the mucus, act on receptor molecules in the cilia membranes. These in turn initiate the cascade of micro-kicks from one signaling molecule to the next to change a membrane protein formed around a tiny channel that lets electric charges flow through it. This alters the electrical potential across the cell membrane, leading to the discharge of impulses in the cell that is conveyed through its long fiber (axon) to the first relay station in the brain.

The narrator's mouthful of crumb-laden tea thus activates a range of receptors tuned to the different volatile components, leading to impulse discharges that carry the information to the brain. But in addition to activating impulses, the signaling cascade in the receptor cells also contains a number of pathways for controlling the sensitivity of the sensory response. Repeated stimulation brings about desensitization of a second messenger pathway. In *Swann's Way*, Proust appears to be describing precisely this effect:

> I drink a second mouthful, in which I find nothing more than in the first, then a third, which gives me rather less than the second. It is time to stop; the potion is losing its magic. It is plain that the truth I am seeking lies not in the cup but in myself. The drink has called it into being, but does not know it, and can only repeat indefinitely, with a progressive diminution of strength, the same message which I cannot interpret, though I hope at least to be able to call it forth again. . . . I put down the cup and examine my own mind.

Desensitization of a sensory response is well known in the experimental literature and is often referred to as *sensory adaptation.* It is a very general phenomenon that occurs any time a given nerve cell or neural pathway is stimulated repeatedly. As discussed in chapter 8, the logic of this is that the nervous system is not constructed to register every sensory stimulus imposing on it, but only those that signal a sudden change from a former state. It is these that carry the most critical information; the same stimulus repeated carries less and less information, until a different stimulus occurs.

It seems clear that desensitization of the initial sensory mechanism takes place during Proust's initial repeated attempts to conjure up the "truth." However, a close reading of the text indicates that several other neural processes are likely also occurring. A second process is adaptation in the neural pathways that process the odor information to give rise to the odor perception; these would occur in the pathways of the brain flavor system. Adaptation may also occur in the pathways that link an odor perception per se to the systems underlying odor memory. Finally, there are the systems related to the narrator's vision of the "truth"—the systems involved in the storage of the visual memories and their retrieval. The fading of the "truth" may thus be due to multiple mechanisms of adaptation, in addition to the desensitization of the receptors. But to assess this further, we need to ask what is the nature of the odor perception elicited by the potion.

———————

Impulses in the fibers from the sensory neurons give rise to spatial patterns of activity within the first brain relay station, the olfactory bulb. These patterns are the smell images of the information carried in the smell molecules (chapters 4–10), which are projected to the olfactory cortex where they form a content addressable memory of the smell object (chapter 11), and are sent from there to the orbitofrontal cortex (chapter 12) to be combined with other sensory and motor systems to form the perception of smell and of flavor (chapters 13–18).

It is this flavor image that was recognized by Proust's brain, at first only indistinctly, as being part of a more complex memory that initially seemed beyond recall. The flavor image of the tea-soaked madeleine is thus metonymic for the complex multisensory image of the town of Combray.

Smell, Emotion, and Memory Recall

The direct access of the smell pathway to these forebrain mechanisms is essential for understanding the nature of Proust's olfactory-evoked experience. We have indicated the cortical mechanisms involved in Proust's cognitive, perceptual response (chapter 18). This direct olfactory connection to the forebrain provides insight into the heightened degree of the emotional state evoked by the odor stimuli, the strength of the voluntary search for the missing "truth," and the overwhelming quality of the "involuntary" memory finally brought forth.

The emotions evoked by the madeleine are central to the whole theoretical edifice of the madeleine episode:

> An exquisite pleasure had invaded my senses, something isolated, detached, with no suggestion of its origin. . . . [T]his new sensation . . . had the effect which love has of filling me with a precious essence. . . . Whence could it have come to me, this all-powerful joy? I sensed that it was connected with the taste of the tea and the cake, but that it infinitely transcended those savours, could not, indeed, be of the same nature. Whence did it come? What did it mean? How could I seize and apprehend it?

Brain research can best provide insight into the question "Whence did it come?" We have explained how, from the olfactory cortex, the pathway for perception is directed toward the prefrontal neocortex. But the olfactory cortex also gives rise to multiple pathways that connect directly to the so-called limbic regions of the brain. These are the phylogenetically old regions of the brain that are involved in the mediation of both memories and emotions.

The key structures include the hippocampus, a central organizing node for single-event "episodic" memories, and the amygdala, which, in parallel with the orbitofrontal cortex, is involved in stimulus reinforcement association learning. The reader may refer to chapters 18 and 19 for an orientation to the places of these structures in the brain flavor system.

Proust believed that the recollection of Combray was an involuntary memory from the past, purer and nobler than memories recalled voluntarily. This notion became favored by a number of literary critics, including the writer Samuel Beckett. From these critics arose what has been called the *salvationist* idea that involuntary memories are kernels of pure truth, the hidden essences of ourselves, precisely because we cannot call them forth at will. But was Combray an involuntary memory? Proust's text says otherwise:

> I begin again to ask myself what it could have been, this unremembered state. . . . I decide to attempt to make it reappear. I retrace my thoughts to the moment at which I drank the first spoonful of tea. I rediscover the same state. . . . I ask my mind to make one further effort, to bring back once more the fleeting sensation. . . . I shut out every obstacle. . . . I compel it [my mind] for a change to enjoy the distraction which I have just denied it . . . for the second time I clear an empty space in front of it; I place in position before my mind's eye the still recent taste of that first mouthful, and I feel something start within me . . . ; I do not know yet what it is, but I can feel it mounting slowly. . . .
>
> Undoubtedly what is thus palpitating in the depths of my being must be the image, the visual memory which, being linked to that taste, is trying to follow it into my conscious mind. . . . Ten times over I must essay the task. . . .
>
> And suddenly the memory revealed itself.

Thus, "the whole of Combray . . . sprang into being . . . from my cup of tea" is hardly accurate, yet many critics seem to have been misled by that dramatic pronouncement. More than a page of text, including the excerpts given here, goes into describing the enormous effort of voluntary recall and searching that took place, and shows that the actual sequence of events involved a considerable delay between the initial sensory input and the realization of the memory associated with it. Proust himself makes clear the delay that is involved, both at the time and much later, in *Time Regained*, when he muses once more on the power of the sensation and the time it took to identify the memory it conjured: "I had continued to savour the taste of the madeleine while I tried to draw into my consciousness whatever it was that it recalled to me."

What are the neural mechanisms responsible for this enormous effort? From a neuroscientific point of view, the association between Proust's intensely emotional state and his highly motivated state is understandable, because much experimental research has shown that the systems responsible for motivated behavior are part of the same deep brain systems involved in emotion. This is summarized in figure 18.2 and the discussion of the human brain flavor system in chapters 18 and 19. The smell and flavor inputs go directly to the forebrain motivational systems that mediated the will to search for the meaning behind the emotions.

Combray Recalled

Our analysis suggests that the recall of Combray is not as involuntary as Proust claims and as literary critics have assumed. From a neuropsychological point of view, the fact that it seems to come back at once and entire is also not surprising; our cognitive mechanisms have a gestalt quality in which we perceive and recall things as integrated wholes. The content-addressable memory mechanism of the olfactory cortex is designed specifically to recall a whole from a small part (chapter 11). Given a fragment, we (and Proust) have a strong tendency to make a whole. Combray comes back in its full visual embodiment in the normal way that most olfactory memories are realized.

Proust seems aware of this phenomenon, and in *Time Regained* he explores in minute detail how one can recognize a person's face when shown only one part of it, such as the nose or mouth. The context for these observations of the "gestalt effect" is a costume party that Proust attends after a long absence from society. He wanders through the crowd, studying the disguised faces that are in effect doubly masked, by costume and by age. Beneath all the costumes are people he has known well yet does not immediately recognize because of his previous absence and because of their disguises, so he methodically studies the face of each person to find some familiar feature that will allow him to compose in his mind the whole face:

> Thanks to a tiny fragment which still survived of the look that I remembered, I was just able to recognise the youth whom I had once met at Mme de Villeparisis's tea-party. . . . I thus succeeded in identifying somebody, by trying to dismiss from my mind the effects of his

disguise and building up, through an effort of memory, a whole familiar face round those features which had remained unaltered.

Proust proceeds around the room observing people and performing this same systematic approach to remembering them, a process that occupies some 40 pages of the novel.

Where in the brain does the memory trace of Combray reside? On the basis of work on other types of memory, we have a good idea that memories are stored in distributed systems in the brain. Given Combray's iconic status, devising experiments to test for this kind of memory has been an irresistible temptation for a number of laboratories that use brain imaging techniques such as positron emission tomography (PET) and functional magnetic resonance imaging (fMRI).

An early study by Gerson Fink and his colleagues in 1996 using PET brain scans tested the idea that there is something special about memories from our own past, the kind of memory that Proust imagined. In this test, subjects sat in the scanner and listened to sentences relating emotional episodes in the life of someone they did not know, while the scanner hummed away recording their brain activity. The results were compared with the activity in subjects who read sentences relating to emotional episodes from their own past. The brain scans showed that the autobiographical episodes activated a specific set of regions. They were primarily in the right cerebral hemisphere, especially in the cortex of the temporal lobe including the hippocampus, parahippocampus, and amygdala as well as the more distant posterior cingulate, insula, and prefrontal areas on the right.

We have already met these regions as parts of the distributed systems for the perception of smell, and for the "images of desire." All these parts of the brain are well known for their involvement in different kinds of memory tasks. Electrical stimulation of many of these areas in human patients by the neurosurgeon Wilder Penfield has long been known to be able to elicit long-term memories from childhood. The PET results suggest that autobiographical memory involves a subset of these regions mainly in the right hemisphere of the brain, the hemisphere commonly associated with more intuitive and nonverbal kinds of behavior.

On the Mechanism of Autobiographical Memory

Jay Gottfried and his colleagues at Northwestern University picked up this question with a study using functional brain scans in 2004. They were intrigued with the fact that an autobiographical memory does not recall just one sense, and that we can usually imagine the whole scene: the sounds of the ocean or the voices of the people there, or the smells in the wind or the food. When we recall a sight or a sound, regions involved in higher-order processing of vision and hearing are reactivated, as if the subjects are internally viewing the memory traces of the original sights and sounds. This is very similar to "mental imagery."

Gottfried and his colleagues asked whether they could demonstrate a recollection in the brain if it involved linking two different senses. They tested subjects by exposing them to a smell (either pleasant or unpleasant) followed by a picture of an object (table, lamp), during which time they were asked to imagine a story or some kind of link between the two. Later, the subjects were placed within the brain scanner, exposed to a variety of objects, and asked which ones reminded them of a smell. Behaviorally, the subjects were best at identifying the association of a visual object with a pleasant smell. The brain activity patterns were dramatic in showing prominent activity in the olfactory cortex, even though the subjects were merely remembering the smell, not actually sniffing it.

How does the brain bind together reactivation of a visual object in the visual cortex with representation of a smell in the olfactory cortex? It was known that in the case of vision and hearing, the memory traces are stored in relation to the different senses but bound together in the hippocampus. Gottfried and colleagues therefore hypothesized that this could also occur in the case of vision and smell. Since the olfactory cortex has no direct connection with the visual cortex, the connection must occur somewhere else; the prime suspect is accordingly the hippocampus and neighboring parahippocampus in the temporal lobe, the same regions found in the study by Fink and his colleagues. The authors conclude:

> [T]he hippocampus binds a distributed trace maintained across sensory-specific regions. Such a system preserves the integrity of the original engram and enables the access by partial or incomplete cues, lending flexibility and adaptability to the memory system. . . . [I]t is less likely

that sensory elements of the original trace would be distributed in higher-order olfactory areas, such as [the orbitofrontal cortex] or cingulate cortex, where sensory fidelity is inevitably compromised through progressive synaptic convergence and divergence. This factor may help to explain the corresponding absence of retrieval-related activity in these particular regions.

This is getting close to telling us what Combray "looked like" in Proust's brain. In these experiments a visual object was used to recall a smell; for Proust, a retronasal smell recalled a visual scene. But the principles appear to be similar: an internal reactivation of the distributed sensory regions bound together by their connections to the hippocampus. Rather than a perfect memory suddenly bursting forth, it appears that Proust was describing positively reinforcing sensory stimuli from a childhood experience, stored in their respective central sensory representations, bound together by their connections to the hippocampus, and reaccessed, beginning with partial flavor cues, as a unified internal image or object by the brain mechanisms of attention, motivation, and emotion.

CHAPTER TWENTY-ONE

Flavor and Obesity

Does knowledge of the human brain flavor system give insights into practical problems like the current epidemic of obesity in worldwide populations? Let us consider the case of fast food.

A "Normal" Meal?

Suppose you've taken a bite of French-fried potatoes. This may not have been the kind of food that Jean Anthelme Brillat-Savarin had in mind when he extolled human flavor, but it is probably as near to a universal food as we have. It was brought to the United States from France by returning soldiers after World War I and gained popularity in the 1920s and 1930s. According to Eric Schlosser, French fries are "the most widely sold foodservice item in the United States." A recent estimate is that 25 percent of vegetables consumed in the United States are French fries. Therefore it might be interesting for a neurogastronome to understand the role of the human brain flavor system in this popularity, especially because it is responsible for many of the culprits leading to obesity.

Sensing flavor starts with the sensory receptors. Your first breath carries the smell through the retronasal route to the olfactory sensory receptors inside the nose. Activation of a combination of the receptors creates in the olfactory bulb the image of the smell that is processed in the olfactory bulb and the olfactory cortex to give rise to the smell perception in the orbitofrontal cortex. This image will dominate your perception of the flavor of the French fries, because it has been learned by you to be a pleasurable smell image, and because smell is the dominant sense of flavor.

At the same time, the French fry will come in contact with the taste buds on your tongue, and you begin to have added to the flavor the taste of the salt on the potatoes. As we have seen, from birth saltiness is built into our taste perceptions as an attractive taste. It is not only essential for life but also enhances smells. Together, a learned smell enhanced by salt begins to be irresistible; as we know, it is hard to eat just one salted peanut. At the same time, we expect a certain sweetness to the potato itself, and the sweetness of sugar is also attractive from birth.

In addition to sensing the taste, the tongue begins to do its motor job, moving the French fries to the teeth for mastication, which engages the motor control of both the tongue and the jaw muscles. The mush that is produced builds up in the space between the teeth and the inside of the cheek, so the tongue keeps sweeping it back to the teeth as well as around the inside of the mouth. The motor control of all these muscles is coordinated so that the movements of the tongue, cheek, and jaw take place almost automatically without our being really conscious of them unless we perceive a taste emerging that causes us to slow and examine more carefully the crushed food. All this motor activity adds to the perception that the flavor comes entirely from the mouth.

Meanwhile, the tongue is sensing the texture of the French fry. This is actually a complex mixture of senses. The crispness is usually one of the most important characteristics, along with the contrast between the crisp outer edges and the soft interior. Sogginess will be unacceptable, as will too hard a crust. There must be a springiness in the "mouth-sense" of the potato to indicate that it is fresh and has been cooked just right. The temperature also has to be right: still warm from the fryer, but not too hot.

Even before eating, we will have judged the French fries by looking at them to be sure the pieces are the right size, with the right color. For example, French fries any color but golden brown would be unacceptable. No one wants to eat a gray or black fry.

As we chew, we expect to hear the familiar sound that starts with a sharp crunching and ends with the soft squish of the mash in our mouths.

Finally, the motor control of swallowing takes over, and the mash goes down to our stomachs (followed by our breathing out for a last enjoyment through retronasal smell).

Three other nonpotato components are extremely important to the flavor: meat flavor, condiments, and companion food and drink.

Consumption of French fries in the United States really took off after World War II as the central item in the rise of fast-food culture. This was based on the fact that, according to Schlosser, potatoes were the food most consumed by Americans after dairy and wheat products. The originators of the fast-food industry figured out that, apart from starting with acceptable potatoes, what really was attractive was the *meat flavor* imparted to them by the oil in which they were fried: "For decades, McDonald's cooked its french fries in a mixture of about 7 percent cottonseed oil and 93 percent beef tallow. The mix gave the fries their unique flavor—and more saturated beef fat per ounce than a McDonald's hamburger." After a public outcry against all this fat and cholesterol, McDonald's changed in 1990 to using a vegetable oil, but with a strong meat flavor due to organic chemical compounds that were produced by the flavor industry. This artificial smell produces the meat flavor that continues to be an attractive part of the French fry's flavor.

In addition to the smell, the flavor industry also contributes other artificial means, through "fats, gums, starches, emulsifiers, and stabilizers" to enhance the mouth-sense (texture) of French fries and other processed foods. Driven by this flavor, a single medium-size bag of French fries delivers 380 calories, almost 20 percent of a day's needs.

The second component is the ketchup. We Americans like ketchup with our French fries. We think of ketchup as being made of tomatoes, but that is only the base. A common brand of ketchup includes the ingredients shown in box 21.1.

You couldn't design a modest sauce to be more flavorful. It stimulates directly three of the five tastes (umami already is stimulated by the meat flavor of the fries). It stimulates retronasal smell with volatiles from tomato concentrate, spices, and onion powder (to say nothing of the unknown artificial compounds hidden under the blanket term *natural flavoring*).

So the ordinary potato is becoming the vehicle for an intense barrage of flavors. A principle of neurogastronomy is that the brain responds to multiple sensory inputs. Satiety to one flavor does not produce satiety to

BOX 21.1
Heinz Tomato Ketchup

Tomato concentrate	Salt
Distilled vinegar	Spice
High-fructose corn syrup	Onion powder
Corn syrup	Natural flavoring

other flavors. As we saw in chapter 9, adaptation to smell occurs in the olfactory cortex, but adaptation to one smell does not affect adaptation to other smells. Similar principles apply to the other senses. This means that we can and will eat much more in response to multiple different smells, tastes, and textures. That property of our brains defines what fast foods deliver, as quickly as possible.

The third component consists of the other foods and drinks we consume with the French fries. For many people, the main course would be a cheeseburger. This adds an even bigger wallop of flavor. The burger, of course, is attractive because of both its orthonasal and retronasal meat aroma. These are enhanced by the fact that frying the meat produces the Maillard reaction, caramelizing the surface of the meat patty to release especially attractive volatiles. The soft chewiness of the meat makes a comforting, familiar contrast with the crunch of the fries. The cheese delivers its own volatiles. The bun gives a soft reassurance to our touch system. A quarter-pound cheeseburger delivers 510 calories.

Some of us like mustard on our burger, which adds the ingredients shown in box 21.2. Again, more taste, smell, and texture are added to our meal.

All this needs to be washed down, usually by a carbonated soft drink. This not only delivers about 10 calories of sugar per ounce (for example, 160 calories for a "medium" 16-ounce [453-gram] can of cola), but also a shower of fizz that stimulates the touch receptors in our mouth and all the way up our retronasal tract to our noses.

The sensory overload, combined with the activation of stretch receptors in our stomachs, leaves us satisfied (satiated, as the psychologists say). In the olfactory bulb, fibers from deep in the brain stem turn off the

BOX 21.2
Grey Poupon Dijon Mustard

Mustard seed	White wine	Pectin
Distilled vinegar	Citric acid	Spices
Salt	Tartaric acid	

mitral cells so that they can no longer be activated by the smell of what we have been eating.

Then it is time to go around the corner to the local coffee shop for an espresso, latte, or cappuccino and maybe a chocolate chip cookie. As we enter, we breathe in the aroma of the coffee by the orthonasal route. As mentioned in chapter 4, coffee contains more than 600 volatile smell-producing types of molecules, all of which flood our olfactory receptors with every retronasal breath as we drink it. A small latte is worth 150 calories; the chocolate chip cookie another 160. The sweet of the cookie and the bitter-sweet combination of the chocolate complement the bitter of the coffee in just the way we have learned to crave.

This overload of flavor is accompanied by an overload of calories. When the calories are added up, the meal just described is, at minimum, more than 1,100 calories, half a day's recommended 2,200, and, at maximum, more than 2,000 calories, almost the recommended minimum for an entire day. The danger to health is obvious. Risks in addition to those for obesity are coming into focus, including those for diabetes and diabetes-related conditions such as insulin resistance, metabolic syndrome, and even increased risks for cancer. The human brain flavor system therefore not only gives us pleasure; it can be hazardous to our health. It is an underappreciated factor in determining the balance between health and disease. Public recognition of this role is needed, as discussed in chapter 26.

Why We Overeat

Knowledge of the brain flavor system can help us understand why we overeat on this menu. First is the sensory overload. The food is high in sensory stimulation and dense in calorie content. A normal diet has more roughage to make us feel full faster, and drinking water with the meal further dilutes the calories, but fast food has too much flavor for too little fill. In addition, we wash down the food with soft drinks that are packed with more calories.

Second, fast food contains a variety of food types and flavors. This is called the *supermarket, smorgasbord*, or *buffet* effect. This idea actually originated with a blind French scientist named Jacque Le Magnen in Paris, who became a legend in research on feeding. In the 1950s he began detailed studies of laboratory rats fed different kinds of diets. He found that on daily lab chow they showed little weight gain, but if he offered them chow with different flavors they quickly began to gain weight. This effect was rediscovered in 1981 by Barbara Rolls and her colleagues at Oxford, who called it *sensory-specific satiety*, meaning that with one flavor the animal quickly becomes full and bored with eating more, whereas a new flavor stimulates renewed eating. This is the effect we all experience at Thanksgiving or buffets or banquets when we feel the urge to go on eating every new dish or course. It is an expression of the fact that the brain is always interested in something new or changing, a characteristic we have seen in all the sensory systems. Although the fast-food industry probably did not know of Le Magnen's research, it designed its foods as if it did.

Another reason people overeat may lie in long-term overstimulation of the skin and membranes of the lips and mouth. These of course are activated by food when we are eating, but, surprisingly, they can also be overactive even when we are not eating. This has been shown by brain scans of activity in the somatosensory area of the cerebral cortex (chapter 13) comparing subjects who are lean and subjects who are obese. As figure 21.1 shows, from the study in 2002 by Gene-Jack Wang and his colleagues at the Brookhaven National Laboratory in New York, obese individuals even in the resting state show higher levels of brain activity in the lip, tongue, and mouth regions of the somatosensory area. The authors speculate that this could reflect hypersensitivity of the receptors there to the rewarding value of food, and could be among the factors associated with overeating.

There are several theories for why we overeat, which have been summarized by Dana Small and her colleagues in a review in 2009.

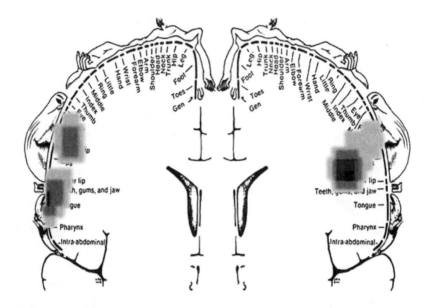

FIGURE 21.1 The human somatosensory system with superimposed intensities of brain activity in subjects at rest
Higher levels of activity (shaded areas) in the lip, tongue, and mouth regions are present in obese individuals compared with lean individuals.
(Adapted from G.-J. Wang et al., Enhanced resting activity of the oral somatosensory cortex in obese subjects, *Neuroreport* 13 [2002]: 1151–1155)

One theory is based on the observation that even though a rat may have fed to satiety, it can be induced by conditioned cues to keep eating. Small and her colleagues mention an experiment that demonstrates this. In this experiment, rats learn to associate the presentation of food with the sound of a buzzer, much like Ivan Pavlov's dogs. If the buzzer sounds when they are sated, they will begin to eat again. Instead of buzzers, humans have many other cues that keep them eating flavorful foods. In our example, a burger cues a bag of potato chips, then the ketchup, then the soft drink, then the . . . This kind of feeding has been shown to be dependent on connections between the amygdala, a node in the emotional network, and the hypothalamus, which is involved in activating feeding. These connections become hypersensitive because of long-standing habits.

Another idea is that overeating is due to ineffective inhibitory circuits in the prefrontal brain regions, combined with heightened excitability of the circuits mediating reward from the foods consumed. An analogy is the involvement of these circuits in drug cravings. This is more evidence that a tendency to overeat involves circuits at the highest cognitive, as well as emotional, levels.

Another factor is the possibility that overeating occurs because eating itself does not have adequate reward value; the brain does not register enough "pleasure" with lesser amounts of food.

In his book *The End of Overeating: Taking Control of the Insatiable American Appetite*, David Kessler, my former dean at the Yale School of Medicine, has emphasized the combination of salt, sugar, and fat as the main flavor villains to be resisted and controlled. Neurogastronomy supports this conclusion, identifying retronasal smell and its associated multisensory brain mechanisms of flavor as underappreciated major factors.

If flavor plays this central role in what we eat, the brain must contain mechanisms for making decisions about whether a food that produces an attractive internal flavor image in the brain is also nutritious. This is a final critical part of the human brain flavor system for determining normal function in healthy people and abnormal function in people who overeat.

CHAPTER TWENTY-TWO

Decisions and the Neuroeconomics of Flavor and Nutrition

The most important ultimate function of the human brain flavor system is making the right choices in consuming healthy or unhealthy food. The key to making these choices lies in the decision-making mechanisms of our brains, which only recently have begun to be recognized. Interest in these mechanisms has merged with the interests of economists, who for many years have realized that people make economic choices that are based on their value judgments about what they like. I first became aware of this when my father, Geoffrey Shepherd, wrote an article about it in 1956. This merging of interests of neuroscientists and economists has given rise to a new field called *neuroeconomics*, a term coined by Paul Glimcher in his book *Decisions, Uncertainty, and the Brain: The Science of Neuroeconomics.*

We have already met elements of the decision system in considering emotions and in the actions of dopamine on the subsystems we have studied thus far (chapter 19), as well as on the factors that lead to obesity (chapter 21). We bring them together here to enable us to enlarge the human brain flavor system to include the mechanisms that determine whether we eat in healthy, and flavor-fulfilling, ways.

Dopamine: Key to a Happy Life

One of the molecules that is key to how our brains work is the neurotransmitter *dopamine*, which was introduced when we discussed emotions in chapter 19. The largest population of dopamine-containing neurons is in the midbrain—far away, you might think, from the highest levels of brain activity. However, from this vantage point these neurons send their axons throughout the brain. Some of them go to the striatum, a large region under the cerebral cortex that is involved in planning, initiating, and carrying out movements as well as in various motivational states; we have already met the striatum in chapter 19 as part of the system of habits for food cravings. Most people have heard about these dopamine cells because of their role in Parkinson's disease, in which the degeneration of these cells and their subsequent loss of input to the striatum results in progressive paralysis.

The other important concentration of dopamine neurons is located in what is called the *ventral tegmental area* (*VTA*). These cells connect widely in the brain. Especially relevant to the flavor system are parts of the striatum, the prefrontal cortex (including the orbitofrontal cortex), the insular cortex (where smell and taste are combined), the nucleus accumbens, the amygdala, and the hippocampus. These VTA–dopamine connections form what is called the *reward system* of the brain. The experiments providing evidence for this have been carried out in rats, monkeys, and humans and often involve the subjects working for rewards of fruit juice, so the subjects essentially are motivated by the images of retronasal smell and flavor.

We have already met Wolfram Schultz in chapter 19 as a leader in studying these dopamine reward systems. In a typical experiment in his laboratory, a monkey explores for hidden food; when a hidden bit of cookie or another food is touched, the dopamine cells release a burst of impulses. In another type of experiment, the dopamine neurons fire a burst of impulses when the monkey is stimulated by the reward of water or fruit juice. The dopamine neurons fire to any rewarding stimulus (discriminating between the stimuli is done by the sensory systems). In his early experiments the cells did not respond to aversive stimuli, such as water that was too salty; the reward needed to be pleasurable. Recently some responses to aversive stimuli have also been found. Of special interest is the fact that dopamine neurons fire to conditioning stimuli, such as a light that signals a future reward. This means they are able to predict future rewards. This ability constitutes one of the highest cognitive functions of the brain. The dopamine neurons do this through their modulation of the cells in the orbitofrontal cortex that are involved in planning future actions. Which brings us back to the human brain flavor system.

These functions modulated by dopamine are important for all sensory systems, but especially so for flavor. The dopamine fibers not only connect from the midbrain to the olfactory cortex, where they can modulate the formation of odor images and odor objects there, but also to the orbitofrontal cortex. In addition, in the olfactory bulb there are dopamine-containing interneurons (some of the periglomerular cells), so that dopamine can be involved in the shaping of the initial smell images in the glomerular layer. Another connection between smell and dopamine is found in the neurodegenerative diseases such as Parkinson's and Alzheimer's; an early sign of these diseases is a decline in smell sensitivity.

Through its role in reward systems in the brain, dopamine is also involved in the brain mechanisms underlying drug addiction. The way it works appears to be as follows. After dopamine is released to activate the reward neurons in the striatum and cerebral cortex, there are cell mechanisms for its reuptake to terminate its action. Cocaine blocks this reuptake, amplifying and prolonging dopamine's action, bringing on the addictive state. Some drugs also increase long-term potentiation at synapses where the excitatory neurotransmitter glutamate is released. Nicotine has been shown to have this effect. It also has a direct stimulatory effect on dopamine cells. There are thus multiple mechanisms for amplifying the reward system. Because of the inherent plasticity induced in brain cells by their activity, these actions tend to be self-prolonging. These addictive effects brought on by drug actions are present in food cravings, as we saw in chapter 19, and were a motivating hypothesis for the early study of food cravings. As we shall see, it is becoming an organizing principle for understanding overeating.

The Reticular System: Your USB Port

Because the dopamine cells have such widespread actions in the brain, it is important to know where they get their inputs. Some of them come from the same areas to which they project, completing feedback loops that can maintain their activity. The other main inputs come from the core of the brain that is often called the *reticular system* because it is not a specific region but rather a kind of network of cells that stretches from the center of the brain stem into the depths of the forebrain. It is an ancient system, present in all vertebrates and expanded in the human. The cells have long dendrites, as if they are reaching out to receive and integrate many inputs from different brain regions. The key point is that their inputs come from within the brain, just as their outputs stay within the brain, so it is an entirely internal system.

This reticular system is the unknown workhorse of the brain. It is rather like the USB slot on your computer, ready to accept a wide range of input devices and connect them to the desired output. The inputs may come from the hypothalamus to stimulate or terminate feeding; they may signal different emotional or motivational states from the prefrontal cortex or the nucleus accumbens, a deep brain region or from different sensory systems, including those involved in flavor. The VTA neurons, with their own long dendrites, integrate these signals and transmit them through release of dopamine, in many cases back onto the same regions that have sent them their inputs. In this way, both the reticular system and the dopamine neurons are concerned with the significance and expectation of sensory inputs or motor outcomes rather than with discrimination among them. The fact that significance and expectation are embedded so deeply in our brains further explains how difficult it is to change the link between flavors and our cravings for them.

Brain Mechanisms for Making Food Choices

We are now in a position to incorporate all the elements of the human brain flavor system into the new field of neuroeconomics. This reflects the fact that economists have realized that the reason people attach economic value to a particular product is to be found not only in the product but even more so in the way an individual places personal value on the product—in essence, the way a person gives it a reward value. An example from recent studies illustrates this new field as it applies to decisions about flavors.

Todd Hare, Colin Camerer, and Antonio Rangel at the California Institute of Technology wished to know how we make choices, and postulated that the brain has mechanisms for making optimal choices between alternatives. It had previously been shown that a value signal for making choices arises in the ventromedial area of the prefrontal cortex, in the frontal lobe, which as we have seen is concerned with higher cognitive functions. They hypothesized that this area must be under control by another area, the dorsolateral prefrontal cortex, which had been shown to be involved in various higher functions, including cognitive control of decision making. They were particularly interested in food choices, and set up experiments to study the brains of people on diets. Tests were first carried out to separate the subjects into two groups, those who demonstrated self-control and those who lacked self-control. The self-controllers chose foods that were healthy, the non-self-controllers chose foods that tended to be unhealthy.

The investigators then put the subjects in a brain scanner and carried out functional brain imaging while the subjects made their choices. They first found that activity in the ventromedial area was correlated with the subject's goal values, whether healthy or not. The activity was correlated with healthy ratings by the self-controllers but not by the non-self-controllers. The dorsolateral area was more active during successful self-control trials. And the dorsolateral and ventromedial areas were both active during self-control trials.

The authors make the interesting suggestion that the ventromedial area originally evolved to assign a short-term value to a food, such as flavor in this case, and the dorsolateral area developed subsequently to reflect long-term considerations, such as healthiness. The dorsolateral area has wide connections with other higher-cognitive brain areas, which the authors suggest may be why general intelligence and emotional control are involved in self-control in decision-making. In final summary, Hare, Camerer, and Rangel observe:

> Lastly, an improved understanding of the neurobiology of self-control in decision-making will have applications to clinical practice in domains such as obesity and addiction, to economic and public policy analysis in problems such as sub-optimal savings and health behaviors, and to legal thinking about which criteria should be used in determining if an individual is in full command of his decision-making faculties and thus accountable to the law.

The Food Choice Control System

A synthesis of the brain systems involved in food choices has been made recently by Nora Volkow, Gene-Jack Wang, and Ruben Baler. Volkow has impeccable credentials for this task. She is a long-time student of drug addiction, as well as director of the National Institute on Drug Abuse (NIDA). As we saw in chapter 19, research on addiction is providing valuable insights into the brain mechanisms that are active in both drug craving and "images of desire" for food. This similarity has led Volkow in recent years to build on this background to propose a model for the different brain systems involved in food choice and healthy versus unhealthy eating (figure 22.1). This puts the basic elements of the model of the human brain flavor system presented in chapter 18 (see figure 18.2) into the more dynamic form of a control system.

The dynamic control model pictured in the figure consists of four main parts. It begins with *saliency*, a psychologist's term for how strong and attractive a sensory stimulus is. Saliency includes the irresistible salty,

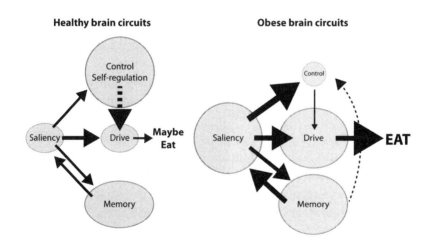

FIGURE 22.1 Schematic representation of the sensory control system in the human
This diagram was developed originally to depict the sensory control system as seen in drug addiction and is here applied to the control of eating.
(Adapted from N. D. Volkow, G.-J. Wang, and R. D. Baler, Reward, dopamine, and the control of food intake: Implications for obesity, *Trends in Cognitive Science* 15 [2011]: 37–45)

sugary, fatty, high-calorie density of fast foods; the smells of coffee and chocolate; and the balanced attractiveness of the flavors of traditional cuisines around the world. These reflect the array of sensory inputs shown in figure 18.2 as well as how much reward value they have as assessed by the brain mechanisms in the orbitofrontal cortex and related areas.

These inputs go to three main subsystems. One is the memory subsystem, which stores the conditioned preferences by learning of the individual. Second is the motivational drive subsystem that determines how much an individual desires or "craves" a kind of food. And third is the subsystem for inhibitory control, emotional regulation, and executive function, providing the top–down cognitive control of choice. Figure 22.1 shows that normally the executive function is strong. In the terms of the study by Hare, Camerer, and Rangel, people with this strong executive function are those with normal self-control.

By contrast, these systems and their interactions appear to be disrupted in obese individuals. As shown in figure 22.1, for them, saliency is powerful, in many cases overpowering, with strong inputs to all the systems. The learned memories of these overly attractive stimuli, the drive they elicit, and the input to the control subsystem are all increased. There is also a new direct input from the learned memories of the craved foods to the control subsystem. But in these individuals, the control subsystem is decreased. As a result, the increased drive from the salient stimuli is only weakly opposed by the inhibitory executive control. There is thus, in the terms of Hare and colleagues, a lack of self-control, and the person experiences a drive for the craved food that cannot be adequately resisted.

This model provides a useful focus for future experiments. The authors point out that other factors are involved, such as circuits that regulate mood and circuits for internal awareness. There is also the highly complex system of regulation of gut hormones, circulating levels of leptins and ghrelins, and other body hormones. And there is the critical role of language in human food choices.

We end by returning to our original question: what is it that makes the flavor of a given food irresistible? Recall the experiments on food cravings that we reviewed in chapter 19. In reviewing those studies in 2009,

Marci Pelchat notes: "Thus, this work supports the common substrate hypothesis for food and drug cravings. The prominent representation of memory and sensory integration structures in this study is consistent with the central role of sensory memory in the experience of food cravings. It is as if, when craving, one has a sensory template of what has to be eaten to satisfy the craving."

This brings us full circle back to the steps along the way we have covered of how the brain creates flavor: the sensory representation of odor images and odor objects—the "sensory template"—in memory circuits in the olfactory pathway; the integration with the other sensory representations in multiple areas of the cerebral cortex; the formation of images of desire by those interacting areas; and the magnification of those desires by activity in emotional circuits beyond the control by the decision-making centers of the brain. The diagrams of the human brain flavor action system (see figure 18.2) and of food addiction control (see figure 22.1) identify some of the hidden brain systems and their mechanisms that need to be taken into account in any public strategy to encourage healthful eating.

CHAPTER TWENTY-THREE

Plasticity in the Human Brain Flavor System

The more we learn about the brain the more we find that it can be changed by activity and experience. At a recent seminar I attended on the subject of critical periods for plasticity of the visual system during development in young animals, the theme was summarized by the speaker in the following way: we traditionally think of the brain as being stable but having limited capacity for being plastic; the new view is that the basic state of brain cells is plastic, and that the brain goes to great lengths to restrain and control this plasticity.

Plasticity applies especially to the flavor system, in two ways: the system contains regions that show a continuing renewal of its cells from stem cells; and experience changes the properties of the cells and their interactions with each other.

Stem Cells Make New Flavor Cells

The brain is born with its full number of nerve cells, except in four regions, where new cells arise from stem cells throughout adult life. It is astonishing that all these regions happen to be key players in creating flavor.

First are the cells of the taste bud (chapter 13). For many years, we have known that they "turn over"; they are born from stem cells at the base of the taste buds, mature to become functional, then die, replaced by new cells. Their life span is around a couple of weeks. It is believed that this turnover is due to the fact that taste buds are exposed directly to all the various kinds of things that come into the mouth. Four-legged animals are particularly liable to take noxious stuff into their mouths, which can kill the taste bud cells, so it seems logical that they should deal with this by a robust mechanism for continual replacement. Another risk that we know from our dear pets is that their habits include licking and ingesting excrement as part of their social interactions—not good for the survival of taste cells.

Second are the receptor cells in the nose (chapter 5). In recent years it has been discovered that they also turn over. These cells are exposed to possible noxious odors with every sniff and inhalation, so it seems logical that they too should be replaceable. In contrast to taste cells, which are entirely contained within the taste bud, the smell receptor cells are true nerve cells, with their receptor-bearing cilia dangling in the mucus of the nose to receive the incoming smell molecules, and their axon fiber stretching all the way to the olfactory bulb. Their turnover rate depends on how noxious their environment is; rats raised in a sterile environment show little turnover, whereas normally it would occur over several weeks. Because we humans keep our noses clean, so to speak, with the help of our erect posture, the turnover rates of our receptor cells must be relatively slow, although this has not yet been established.

Third are the small cells, the interneurons—periglomerular cells and granule cells—in the olfactory bulb that shape the processing of smell signals through their inhibitory actions on the larger mitral and tufted cells (chapters 7 and 10). Recent work, also in animals, has shown that the interneurons also undergo turnover. This occurs through stem cells that are located not in the olfactory bulb but far away in the base of the brain, among the stem cells that give rise to the cells of the cerebral cortex. The cells that give rise to the cerebral cortex stop making new cells before birth, but others form a "rostral migratory stream" that moves along a long narrow path forward and into the olfactory bulb, to become integrated into the populations of periglomerular and granule cells. Given all the current interest in stem cells within the adult brain, these olfactory bulb cells have become the subject of intense investigation.

With its input and interneurons constantly turning over, the only stable players in the olfactory bulb are the large neurons, the mitral and tufted cells. This makes the olfactory bulb, the region that plays a central role in flavor, one of the most plastic regions in the entire brain. Is this turn-over needed to replace cells lost to infection? Or is it needed to deal with the unique qualities of the odor stimuli? We require much more research to answer this important question.

The fourth region in the brain that shows turnover of new nerve cells is the hippocampus, specifically the region called the *dentate gyrus*. The hippocampus is the portion of the brain that is essential to learning and memory for all the senses, including smell and flavor. We saw how, in the example of Combray, it is the region that connects the memories of smells, sights, and sounds that are stored in their different central brain systems and binds them together to produce the unified memory containing all of them (chapter 20). Why this function should require continual replenish-ment of new cells is also unknown.

Plasticity in the Taste System

The other main kind of brain plasticity is evidenced in changes in cell properties induced by their activity. Suspicion immediately falls on the regions of high plasticity because of the cell turnover. We can begin with the taste bud.

Linda Kennedy and her colleagues at Clark University asked the fol-lowing question: If we eat a lot of sugary foods, do we change our sensi-tivity to sugar and our sense of sweetness? They first discovered among a panel of subjects that some of them had relatively low sensitivities to sweet taste as tested by glucose. So they wondered if they could raise the subjects' sensitivity, and indeed found that with further repeated testing with sugar the sensitivity—including the ability to discriminate the sweet taste—increased. They suggested that this must be due to an "experience inducible mechanism," which suggested a more general "human taste induction" hypothesis that repeated exposure to a taste stimulus in-creases sensitivity to it.

To test this further, they carried out a study in which they used re-peated exposure of subjects to fructose, a slightly different sugar that tastes twice as sweet as glucose, and also tested with glucose. The fruc-tose did indeed increase the sensitivity to glucose. This showed that the mechanism was relatively general, not narrowly focused on just one kind of sugar molecule. They also showed that with just five brief exposures to glucose over as many minutes for a few days, the sensitization lasted for almost two weeks. After testing stopped, the sensitivity returned to normal over several weeks. This means that the effect is reversible, and that the reversibility is evidence of the plasticity of a normal response. Other experiments have shown that higher sensitivity can be induced by other taste stimuli, such as to monosodium glutamate, which causes the umami taste, and the substance glutaraldehyde.

Where is this effect taking place? An obvious place is the taste bud, with its vigorous turnover of cells. Maybe the new young cells are more sensitive to rising concentrations of the sugar than the older cells. Perhaps with the cessation of exposure to high sugar content, they are more sensitive in adjusting their responses to the lower levels. Evidence that the change in sensitivity is occurring in the taste bud comes from animal experiments in which the impulses in the taste nerve, the chorda tympani, are recorded, showing that with repeated stimulation the responses increase. However, changes can also be occurring centrally. Functional brain imaging has revealed intensifying responses in central parts of the system—for example, in the cortical taste area. Further experiments are required to distinguish between the two levels.

There have been few follow-up studies and little public comment on these findings. However, if repeatedly eating foods high in sugar increases the sensitivity to sugar, nutritionists should be alerted to the fact. With the commercial soft-drink industry pushing ever larger amounts of sugar-laden drinks at the consumer, it is not unreasonable to suggest that the "human taste induction hypothesis" is being confirmed on a scale of millions of subjects.

Plasticity in the Smell System

In the nose, the turnover of olfactory receptor cells has suggested that they might also be sensitive to repeated stimulation. Investigators at the Monell Chemical Senses Center in Philadelphia tested this idea with several types of smell molecules. They started with the pheromone molecule androstenone, which is secreted in the sweat and urine of males and can be detected by many females in whom it can elicit sex-related behavioral responses. The investigators found that repeated exposure could raise sensitivity and induce the ability to detect androstenone even in males. Extension of these experiments showed that sensitivity in humans to other kinds of odors, such as citralva, could also be increased by repeated exposure.

Where might this increase be occurring? In experiments with both humans and animals, evidence was obtained that increased sensitivity occurs in the receptor cells in the nose. What about a central effect? If an increase occurs in the receptor cells, it would be difficult to find evidence for a separate increase mechanism more centrally in the brain. However, Noam Sobel and his colleagues at the University of California, Berkeley, in 2002 devised a clever experiment to get around this problem. They delivered androstenone to one nostril only, while the other was blocked, and then tested for sensitivity to androstenone in the previously blocked nostril. The sensitivity was shown to be increased. Therefore it must have been due to changes in the central pathway.

These experiments involved only orthonasal sniffing. One can hypothesize that the same changes in sensitivity may apply also to retronasal smell related to flavor. The more we sense the smell component of a flavor, the more sensitive we may become to it. This, too, should be of interest to nutritionists.

Learning to Smell

What kind of central mechanisms might be involved in experience-dependent changes in sensitivities to smells and tastes? One type of mechanism may be similar to the effect of drugs of abuse in altering the actual cellular properties of nerve cells, as we discussed regarding "images of desire." Whether flavor signals can have such strong effects is under investigation.

One of the leading ideas for the mechanism of experience-dependent changes underlying learning and memory goes back to a psychologist, Donald Hebb. In 1949, he published *The Organization of Behavior*, in which he suggested that when a neuron carrying a signal causes an increase in signaling of another neuron to which it is synaptically connected, the two intensify their connection to make it more effective with further activation. In the 1970s, this kind of connection was found in the hippocampus, in which stimulating one set of neurons caused an enhancement in their connection with another set. This accordingly was called *long-term potentiation*, commonly known as *LTP*.

Long-term potentiation can be demonstrated in many places in the central nervous system. In addition to the hippocampus are the olfactory cortex and the neocortex. The microcircuits we have described in these regions are therefore prime candidates for mediating central changes in sensitivities to smell and taste stimuli. This view is expressed very well by Donald Wilson and Richard Stevenson; in their book *Learning to Smell: Olfactory Perception from Neurobiology to Behavior*, summarizing their work on learning and memory in the olfactory cortex, they conclude: "The most striking characteristic of human odor perception is its plasticity."

This plasticity is expressed not just as changes in cellular properties, but also in the ways that distributed systems can shift to shape the sensory responses. Wilson and Stevenson go on to explain that experience can change our actual perception of a smell, how it differs from other smells, and whether we are attracted to it or repelled by it. They emphasize that this context for smell perception depends on the formation within the smell pathway of a perceived "smell object," in the way we described in chapter 11.

In fact, all sensory and motor systems have the potential for these changes. As indicated in the opening paragraphs, we now regard the nervous system as a highly plastic slate on which experience writes its images and memories. Since these images and memories change the slate, perhaps Brillat-Savarin had the right idea when he wrote: "Tell me what you eat, and I will tell you what you are."

From this perspective of the plastic brain, we can give poets their due in having anticipated the neuroscientists. In "The Dry Salvages," the third poem in his *Four Quartets*, T. S. Eliot wrote:

For most of us, there is only the unattended
Moment, the moment in and out of time,
The distraction fit, lost in a shaft of sunlight,

The wild thyme unseen, or the winter lightning
Or the waterfall, or music heard so deeply
That it is not heard at all, but you are the music
While the music lasts.

To which we may add the paraphrase:

Flavor sensed so deeply that it is not sensed at all,
But you are the flavor while the flavor lasts.

CHAPTER TWENTY-FOUR

Smell, Flavor, and Language

A *New York Times* story about a dessert prepared by the French actor Gérard Depardieu brings together several themes of neurogastronomy:

> "The other day we made a fondant of apples with three different kinds of apples—Canada, Granny Smith and Calville," Depardieu said. "We did a white caramel sauce, put that on the plate. Then in the compote of apples, slightly reduced, I put a bit of butter and an egg yolk, and put that in the oven. It's not at all like a compote because it's lightly gratineed on top. We served it with a creme Anglaise and a caramel ice cream, and I said: 'It's far too sweet, too rich. Nothing explodes, there is no subtlety.' So what we did is put an apple sorbet and a drop of rum—old rum that Fidel Castro took from the cave of Battista 50 years ago. He gave me a few bottles. And that"—he paused with a theatrical flourish —"*that* is God in velvet underwear."

The quote shows the central role that flavor plays in our lives. It shows how fascinated people are by the procedures of cooking, one of our defining human characteristics. No other animal would be so crazy as to attempt to combine all these elements just to have a meal. There are at least 10 ingredients in this little dessert dish, ranging from raw apples to 50-year-old rum. They are subjected to several methods of preparation, from freezing ice cream and sorbet to heating a compote in the oven. There is testing at each stage all along the way for just the right flavor combinations.

And there is language. The dessert could not have emerged without language: words to describe and discuss each flavor combination at each stage; words to express the disappointment with the excessive sweetness and lack of subtlety and to find the solution; and finally, words to express the joy at the right result.

Human Language Is Essential for Human Flavor

In the action section of the human brain flavor system (see figure 18.2), language is therefore an essential part. It is one of the main reasons our flavor system is unique among all animals. If cooking is the defining characteristic of humans, as James Boswell averred, flavor is its joy and language is its handmaiden.

This essential role of language in relation to smell and flavor may come as a surprise. The received wisdom (I have believed it myself) is that we are poor at describing our world of smells. There are no words that characterize smells in specific ways, the way we use words to describe colors or simple shapes, for example. Language seems to fail, to resort to analogy, metaphor, or reference to the object that produces the smell. Thus, things smell "sweet," like the taste of sugar, or, for the scent of bananas, "like bananas." For an organic chemist, there may be meaning in describing the scent of bananas as the smell of amyl acetate, but if you ask how amyl acetate smells the answer will probably be "like bananas."

The argument against a close link between smell and language is also made on scientific grounds. As we have seen, the cortical receiving area for smell is in the most anterior part of the frontal lobe. By contrast, the cortical area for the motor control of language is Broca's area in the posterior part of this lobe, and the area for the sensory integration of language, Wernicke's area, is in the temporal lobe even further away (see figure 13.1). This shows, it is argued, the tenuous connection smell perception has to language. A similar argument would apply to the relation between flavor and language.

Wernicke's area is close to the auditory receiving area, which seems logical because of its direct involvement in interpreting the sounds of the spoken language, and Broca's area is close to the motor area for control of the muscles of speech. Apart from that, the two language areas are related by long-distance connections to many parts of the cerebral cortex in order to form the spoken language in the service of all the sensory and motor systems of the brain. Just as the frontal lobes are connected to the visual areas in the occipital cortex, so are the prefrontal areas connected to distant parts of the cortex, including areas related to the speech areas. What matters is whether the connections can subserve not only the sensory functions related to flavor but also the learning required in relating the perceptions to the neural machinery of language.

There are strong indications that they can. We have already seen that the ingestion of food, and the retronasal smells involved in flavor, occur through the mouth, the same orifice used in the production of language. Food and language are not only close neighbors in this sense; they occupy the same house. Further reasons for a close link between smell and language are, first, the strong evolutionary link between the rise of cooking and the rise of language, and second, the extraordinary vocabulary we actually have developed for flavors (witness the Depardieu dessert).

From the earliest documents of recorded history, it is obvious that foods and their preparation have been high priorities of human societies. It is from these times that the traditional cuisines from around the world have emerged.

In perusing some of the history of those times from the perspective of the sense of smell and its role in food flavors, I have been particularly struck by the richness of the cuisines and the associated richness of the cuisine vocabularies. For example, by Roman times there was a huge demand for herbs and spices imported from Arab traders, who brought them through the silk and spice routes that extended from the Far East (this was a thousand years before Marco Polo). In *The Spice Trade of the Roman Empire*, J. Innes Miller lists the spices that were mentioned in the cookbooks of the time; they totaled 87. By comparison, a recent edition of *The Joy of Cooking* lists only 18. This vast array of herbs and spices, each with a name and a characteristic aroma and flavor, indicates the richness of both the cuisines and the languages to describe them from earliest times.

The Vocabulary of Flavor

Today, the vocabulary of food aromas and flavors seems overwhelming. It would help if we could classify all smells and flavors as arising from the mixing of a few "primary" smells, in analogy with how colors arise from just three primary colors, but smell and flavor are too complex. There are in fact several vocabularies. Organic chemists have produced thousands of terms for the smells and flavors of the molecules they synthesize. Psychophysicists have their vocabulary for the flavor components isolated by mass spectrometry, as described in chapter 4. And food scientists have their terms for characterizing the flavors of foods, as do wine tasters for their wines.

An example may be found on the Flavornet Web site, maintained by the leading psychophysicist Terry Acree and his colleagues at Cornell University: a compilation of more than 700 odorous food components, with data on their chemistry and sensory properties, organized into 25 classes, such as fruity, cooked meat, dairy, fishy, spicy, and so on.

This list provides dramatic evidence that the large universe of smell molecules produces a corresponding large universe of smell perceptions and the vocabulary to give them identities. This is amplified by the associations each of the food smells has with the flavor of which it is a part.

The intriguing idea is that all this vocabulary plus the syntax and grammar to communicate it reflect the attempt on the part of humans to describe their world of smell and flavor. Some claim that humans can discriminate among 10,000 odors. However many it is, there must be a corresponding number of words to describe them. Moreover, we have seen that the sense of smell involves not only the perception of a scent, but also the associated memories that are evoked and the emotions that are attached to it. These amplify the vocabulary, so that as we saw with the incident of Marcel Proust and the madeleine, a perception can bring back the whole scene of a bygone time and the emotions connected with it, all requiring the use of language in order to identify the memory, describe the emotion, and communicate it to others.

A major hypothesis of this book is that the reason it is so difficult to describe these smell and flavor perceptions in words may be that *they are represented in the brain as arbitrary irregular patterns of activity, what we have called "smell images."* As we argued in chapter 8, it is difficult to describe in words a nongeometrical visual image such as a face, even though we can identify it unerringly. In like manner, it may be postulated that it is difficult to describe in words a smell image, even though we can identify it unerringly as well.

Thus, connecting smells and flavor with language may be difficult, but it is a uniquely human endeavor. That we require effort to do it, using all the linguistic tricks at our disposal (analogies, metaphors, similes, metonyms, and figures of speech) qualified by the entire vocabulary of emotion (joy, despair, hate, revulsion, craving, and love) should not come, therefore, as a surprise. Gérard Depardieu was only doing what came naturally.

Among the best challenges to the use of language to evaluate flavors is wine-tasting by experts. In chapter 15, we saw the importance of language in assessing the flavors of red and white wines when complicated by the factor of color. Here are two further examples of using language to characterize wine flavors.

Ann Noble and the Wine Aroma Wheel

Ann Noble is such an expert, having devoted her life to the scientific analysis of wine tasting in her laboratory at the University of California, Davis. I visited Noble in her home in Davis several years ago, and also her laboratory for wine tasting. For this work she has developed "The Wine Aroma Wheel," with terms to describe wines organized in three concentric circles, starting at the center with the most general terms (*fruity, earthy,* and so on), to more specific (*berry, citrus,* and the like), to the most specific in the outer circle (*blueberry,* other specific fruits, and flavors). She proposes that this provides a logical system for using language tags to work one's way through a hierarchy of classification of percepts. It enables even the beginner to recognize varietals (based on the type of grape) and the "notes" of different wines.

Robert Parker and the Language of Wine

Robert Parker, renowned for his million-dollar nose, stands in contrast. In his *Parker's Wine Buyer's Guide* (from my 1995 edition), he catalogues more than 7,500 wines, based on his ability to discriminate among different labels and different years. How does he do it? Let us look at some examples of his descriptions.

His summary of the qualities of one of the great wines, a 1994 Petrus Bordeaux, starts with an overall comment on it being a "powerful, tannic, backward looking wine." *Powerful* presumably means strong-tasting, *tannic* refers to the astringency due to the binding of proteins (chapter 14), and *backward looking* presumably means a long aftertaste. The color he describes as "deep, dark ruby/purple," the darkness increasing with age and complexity. He then goes on to comment that "with coaxing" (slow careful swishing of the wine in the mouth?), the "closed nose [presumably meaning retronasal smell with the wine held in the mouth] offers up scents of coffee, herb-tinged, jammy black cherries, and toasty new oak." On his scale of 100 points being best, he rates it a 91–93-plus.

For comparison, the next wine in the list is a 1994 Phélan-Ségur (Saint-Estèphe), which has an "impressive saturated purple color, sweet, jammy cassis aromas, medium body, moderate tannin, and fine ripeness and chewiness in the mouth and finish." Here we have almost all the senses combined in the evaluation: purple color, sweet taste, cassis aromas, and tannin feel, plus even the motor quality of "chewiness" in the mouth and finish. From this perspective one can begin to see the logic in Parker's approach, using language to differentiate the qualities of all the sensations and motor qualities of the wine. In addition he notes the "finish," which includes the dimension of time for the evolution of the whole effect.

For comparison, let's go to California, where Parker has built up enthusiasm for wines that stand equal to the best of France. He especially likes the Robert Mondavi wines from the Napa Valley. Among the choicest is the Cabernet Sauvignon Reserve. He describes the 1990 vintage as exhibiting an "opaque dark ruby/purple color." There is a "huge" nose, full of "abundant aromas of sweet, toasty, new oak, jammy cassis, and roasted nuts," with "a concentration of cassis fruit welded to sweet, soft tannins." The finish is "velvety until some tannins emerge as the wine sits in the glass." Here again we see the evaluation emerge from a combination of color, taste, nose and mouth-feel, and the evolution of the flavor with time, not only from the lingering flavor from what is left in the mouth and throat but also from the wine simply sitting in the glass.

In 2000, Frèdèric Brochet and Denis Dubourdieu in Bordeaux analyzed the language used by Parker and three other wine experts. They found that the specific sensory qualities were evaluated within the context of whether the taster liked those properties, singly or together. This is the same axis along which Noam Sobel and his colleagues arrange smells (chapter 12). Brochet and Dubourdieu conclude:

Although many efforts have been made to characterize the quality and flavor of the compounds in wine by gas chromatography and other chemical techniques, tasting remains the single universal test used to assess properly wine sensory properties.... The main cognitive concern regarding flavors is whether they are good or not. This concern is so strong that even experts cannot ignore it and it is what drives the organization of their descriptive language. In this way experts are not so different from novices.... P[arker] evaluates as would a novice, and this could be one reason for the extraordinary success of this writer.

Language Is Difficult for Faces, Art, and Music, Too

Because the concepts of smell images and flavor images, as well as images of desire, are so new, it is understandable that there have been no studies on how to describe a wine in terms of these images. We can therefore only speculate. If we recall that the core problem is describing a highly irregular smell image, it may help to compare this with the equivalent problem in other sensory systems. We think we can use language in very precise ways in vision and hearing, but is this true? Surprisingly, I think not, if we use our visual analogy again.

In vision, familiar objects—tables, houses, flowers—are so easily described that it seems that this is the system best connected to language. However, is this true for irregular objects? For example, a face is a complex irregular spatial object. As noted in chapter 8, humans are extremely good at recognizing faces. This ability is of critical adaptive value in recognizing not only a particular individual but also the nuances of facial expressions, conveying all the emotions that enter into human relations.

Recognizing a face, as we have discussed, is a case of pattern recognition. It is often claimed that this is one of the things humans are best at. We've already indicated how, if you are shown a number of pictures of grandmothers' faces, you will have no difficulty picking your own grandmother out very quickly. But can you *describe* to someone else in words how you identify her? Try to describe for yourself your grandmother's face in a way that would tell others exactly what she looks like. You might agree that we are at loss for words to describe a face, just as most of us (although apparently not Robert Parker) are at loss for words to describe a wine.

Another example from vision is painting. Representational art is easy; describing in words a painting of a landscape by Constable or Cézanne or a sunflower by van Gogh can be done by anyone. However, Picasso and the cubists present more difficulty, and current modern painters even more so.

Take, for example, Cy Twombly. An American artist born in 1928, he developed as an Abstract Expressionist whose influences came from Europe, Africa, classical art, and mythology. Many of his works contain differently colored scratchy blobs seemingly scattered over the canvas, with, however, a disciplined structure that is impenetrable to the casual viewer. Here is a description of his style from a recent exhibition:

> Twombly's "literariness" is something that has consistently told against him, along with his fancy foreign ways and his "insinuating elegance." But his art, as [Nicholas] Serota acknowledges in the catalogue, has always been elusive and, for many people, even enthusiasts of contemporary art, unfathomable. Twombly himself has maintained an unusual reticence. In the mid-50s, he wrote a short statement for the Italian art journal *L'Esperienza moderna*: "To paint involves a certain crisis, or at least a crucial moment of sensation or release; and by crisis it should by no means be limited to a morbid state, but could just as well be one ecstatic impulse."

Literariness, insinuating elegance, and *unfathomable* are words that would be in some people's wine vocabularies. It appears that observing the complex visual image of a vintage Twombly brings forth the same challenges to language as sensing the flavor image of a vintage wine. This is exactly what was implied when introducing the concept of smell images in chapter 9.

Another example comes from the world of music. We can all recognize and reproduce a simple melody. However, classical and jazz composers seek much more. Bruno Walter (1876–1962) conducted many of the great symphony orchestras around the world in the first half of the twentieth century. In his memoir *Of Music and Music Making,* he describes the music he conducted: "Music springs from and is replenished by a hidden source which lies outside the world or reality. Music ever spoke to me of a mysterious world beyond, which moved my heart deeply and eloquently intimated its transcendental nature."

Here, too, a wine connoisseur could have used some of those same words. It appears that the complex temporal images of music are as difficult to describe in words as the complex spatial images of art and the complex brain images of wine. A picture created within the brain is worth a thousand words, whether it is representing an abstract painting, chords of music, or the molecules of smell.

CHAPTER TWENTY-FIVE
Smell, Flavor, and Consciousness

How much of flavor perception is conscious and how much unconscious? Studies of smell are beginning to provide some intriguing answers. Traditionally, the neural basis of consciousness was ruled out as a subject for scientific study. However, recently it has become a growth industry in neuroscience. Leading the way have been a computational neuroscientist, Christof Koch, and his colleague Francis Crick. Crick, with James Watson, solved the structure of DNA in the early 1950s. He then turned his interest to neuroscience. Beginning in the 1970s, he took on the question of the neural basis of consciousness as a special interest, elevating it to one of the key problems that needed to be solved to understand the human brain and mind. Most neuroscientists, however, regarded it as an "ill-posed" question: *consciousness* was too vague a term to be the subject of scientific experiment. Most still think so.

Nonetheless, the two forged on. In a well-known synthesis in an article in 2003, Crick and Koch first define "consciousness" in a relatively narrow sense, only for the visual system, as "perceiving the specific color, shape or movement of an object." They outline a framework in which primary sensory cortex at the "back" of the brain contains cells and microcircuits that act as "feature detectors" of the information relayed from the thalamus. Feature detection is believed to be largely unconscious. According to Crick and Koch, "The conscious mode for vision depends largely on the early visual areas (beyond V1) and especially on the ventral stream [in the temporal lobe]." This response is then relayed to the "front" of the brain, followed by complex backward and forward interactions between the prefrontal cortex and the visual association areas. One suggestion is that the forward connections are largely "driving" and the backward connections are largely "modulatory." Consciousness, in this view, arises from special (as yet unspecified) firing properties of cortical neurons, in particular those that project from the sensory association areas to integrative (not primarily motor or sensory) areas of the prefrontal cortex.

No indication is given of how olfactory "conscious" perception would fit into this scenario. But smell and flavor raise many interesting questions about consciousness. In olfactory perception there is no "back" of the brain; the primary neocortical receptive area is in the orbitofrontal cortex, which is in the prefrontal area. Thus, in olfaction, all the sequences of processing necessary to get from the back to the front of the brain are compressed within the front of the brain itself. This reflects the evolutionary position of smell, with its privileged input to the highest centers of the frontal lobe throughout the evolution of the vertebrate brain. From this perspective, the best chance to reveal the neural basis of consciousness in mammals, including primates, should be sought in the olfactory system and its role in flavor.

"Is There Such a Thing as Blind Smell?"

These considerations were furthest from my mind a few years ago as I was getting ready to give a talk at the Society for Neuroscience on our first computer models of the interaction between an odor molecule and an olfactory receptor model, when Crick and Koch came in and took seats. I was elated with the prospect of having all that wisdom from the discovery of DNA applied now to how an odor molecule acts within the binding pocket of an olfactory receptor molecule. Sure enough, in the question period Francis rose, and I waited expectantly to hear his words of molecular wisdom: "Gordon," he said (here it comes, I thought), "I wonder if you can tell me: Is there such a thing as blind smell?"

I was so surprised by this unexpected question that I could only sputter that I didn't know, although I understood right away what he meant. He wanted to know if smell has something similar to a puzzling finding in vision called *blind sight*. This is demonstrated in patients who are legally blind from an injury to the visual cortex; if you show them pictures, they say they do not see anything. However, if they are forced to answer questions about them, the answers are above chance. It is as if they have rudimentary sight although they are blind; hence, "blind sight." It indicates that without being conscious of it, their brains, containing a visual pathway rising only to the level of the thalamus, can actually register the visual scene.

Odors that we sniff but do not perceive? We never did resolve the question before Crick died several years ago. Of course, a pheromone in some ways qualifies as a "blind smell," because it stimulates the olfactory pathway to bring about a behavioral response without the animal being conscious of it as a perception. Noam Sobel and his colleagues at the University of California, Berkeley, showed, using stimulation with a supposed human pheromone, that activation of fMRI signals in the brain occurred without awareness of stimulation by the subjects. But what Crick was interested in was not pheromones, but rather a situation in which a person has a brain injury that renders him or her unable to perceive smell stimuli consciously, yet can be shown to be able to perceive them unconsciously. The question goes to the heart of how conscious smell perception—and with it, flavor perception—arises in a pathway that does not go through the thalamus. There is now an answer. But first a little background.

Some Clues from the Smell Pathway

For a sensory stimulus to be perceived, we need to be conscious, alert, and paying attention. Understanding all these brain mechanisms would take us far beyond the main subject matter of neurogastronomy. However, the question of consciousness is relevant, because we cannot make use of our internal flavor image unless we are conscious. Thanks to Crick and Koch, the neural basis of consciousness is now a problem attracting some interest in neuroscience as well as in psychology and philosophy. Almost all the evidence and ideas relate to the visual system. The olfactory system has been almost entirely ignored. But, building on the facts we have covered, it may hold some interesting insights.

Where in the smell pathway does conscious perception of smell arise? Let us see if we can find any clues from what we have learned.

To review, the first smell station in the brain is the olfactory bulb, where the odor image is first formed and initial processing takes place. This processing is subject to the behavioral state of the animal through fibers originating deep within the brain. These fibers modulate processing of the odor image there depending on whether we are hungry or full (chapter 10). They also carry information about whether we are waking or sleeping, which is quite relevant to whether we are conscious or not, but it appears that this modulation occurs more centrally in the brain. We conclude that conscious perception of smell and of the spatial smell patterns does not arise in the olfactory bulb.

The next station is the olfactory cortex. Many regard this as the "primary" olfactory cortex, but, as discussed in chapter 12, in other sensory systems the term *primary* is reserved for the first area that the thalamus connects to in the neocortex. For example, in the visual pathway, conscious perception requires interactions between the thalamus and V1, its projection area in the occipital lobe at the back of the brain. Lack of V1 is what makes blind sight so paradoxical. It appears that there are a few thalamic fibers that project to surrounding association visual cortical areas to mediate this "unconscious" sight. Normally these surrounding areas elaborate the simple response in V1, but if they do not get their input from V1, conscious perception does not occur.

Regardless of whether conscious smell first arises in the olfactory cortex, the smell pathway, like all other sensory pathways, continues to the orbitofrontal cortex. I prefer to call this connection to the neocortex the *primary olfactory cortex*, with the use of *primary* similar to its use in the other sensory systems. In analogy with V1 in the visual system, it could be called "O1" in the olfactory system.

The connection to O1 of the orbitofrontal cortex is both direct and indirect (chapter 12). The larger direct projection is carried by the output axons of the pyramidal cells. These are the same axons that give rise to lateral inhibition within the olfactory cortex and the long re-excitatory association fibers that also carry centrifugal information back to the olfactory bulb. The smaller indirect projection is through activation, by the pyramidal cells, of cells in the endopiriform nucleus just deep to the pyriform cortex; their axons project to the mediodorsal thalamus, where they synapse with cells that also converge onto the olfactory orbitofrontal area.

Conscious Smell Perception

A big challenge is to understand the special nature of the direct projection that does not pass through the thalamus, a feature that makes the olfactory pathway unique among sensory systems. Surprisingly little attention has been given to this remarkable feature. The implications for conscious sensory perception should be one of the most intriguing challenges in cognitive neuroscience.

There are two possibilities. One is that conscious smell perception arises already at the level of the olfactory cortex. This idea is supported by the finding by Verity Brown in 2007 at the University of St. Andrews in Scotland that rodents in which the olfactory area of the orbitofrontal cortex has been removed can still perform normally on an odor identification task. If behaving rodents are regarded as "conscious," this finding is significant for two reasons. First, it means that this is the only sensory pathway that gives rise to conscious sensory perception without reaching the neocortex and without engaging the thalamocortical system. Second, if this is so, we must ask: What are the subcortical mechanisms that replace the thalamus and the neocortex? In other sensory systems, the thalamus is involved in relaying brain stem activity that subserves "arousal" of the whole cortex when we are awake. In addition, Crick suggested many years ago that the thalamus also acts as a kind of attention "searchlight" to direct consciousness to the stimuli that need to be attended to. Because olfaction apparently does not have these mechanisms, how are the pathways for smell and for flavor coordinated with these thalamocortical mechanisms so that our perception appears to be conscious for smell in the same way as it is for the other systems?

These findings may apply particularly to rodents, because the same experiments in dogs produce profound deficits in the ability to smell. According to Noam Sobel and his colleagues, human subjects with trauma to the right thalamus show deficits in the ability to identify smells, and they also were found to experience less pleasantness of pleasant smells;

this suggests that in humans the neocortical level is as necessary for smell perception as in other sensory systems, a finding that must generalize to flavor perception as well.

Answering Crick's Question

This evidence in humans has received support from an unusual patient reported in 2010 by Jay Gottfried at Northwestern University. A 36-year-old patient named S. came to the hospital after suffering an injury to his head on the right side while falling down a flight of stairs. The injury caused bleeding into his right frontal lobe. His recovery proceeded well, except that he experienced a complete loss of his sense of smell, a condition called *anosmia*. A functional brain scan showed that he had damaged his right orbitofrontal cortex; other olfactory structures were not affected. Gottfried heard about the case and realized that it presented an unusual opportunity to test the role of the right orbitofrontal cortex in conscious smell, and brought S. to the laboratory for testing.

The subject's psychological state was judged to be normal. Gottfried and his colleagues employed a standard method using *Sniffin' Sticks*, small paddles coated with odor beads, to test (orthonasally) for smell perception. The test confirmed that S. was unable to detect any odor in either nostril, even at strong concentrations. The fact that the injury was only to the right orbitofrontal cortex supported the idea that most conscious processing of smell at the neocortical level occurs in the right orbitofrontal cortex. This kind of "lateralization" is unusual in sensory systems.

Even though S. was not conscious of the odors, he showed the ability to detect odors on the left side (the uninjured side) when tested against odorless controls. This meant that the left side smell pathway appeared normal whereas the right side was nonfunctional. Correlated with this, brain scans showed activity in the left olfactory pathway—the olfactory cortex and the orbitofrontal cortex—and the amygdala, often implicated in higher smell processing. The brain scans also showed activity in the right pathway up to the olfactory cortex, but none in the region of the damaged right orbitofrontal cortex. Finally, galvanic skin responses (the "lie detector" test) were used; they showed emotional responses to both right and left nostril stimulation—a further example of a response on the left side that was not consciously perceived.

Control experiments were carried out in normal uninjured subjects, which showed normal smell perception in both nostrils. The brain scans showed more activity in the right-sided smell pathway, supporting further the idea that smell processing tends to be lateralized to the right. It is interesting of course that the right side of the brain is believed to be the less logical and more artistic in its functions, which might be considered to be appropriate for elaboration perceptions of smell. The authors conclude that

> [t]he fact that left-sided odor stimulation in Patient S. elicited appropriate peripheral and central responses suggests that the left olfactory system was largely preserved to support processing of the perceptual and emotional content of an odor, yet was unable to assign conscious awareness or feeling to that odor. Taken together, our data provide some of the first evidence to support the central role of the right OFC in facilitating the transformation of an upstream olfactory message into a conscious percept.

And what about Crick's question? Gottfried and his colleagues conclude that "these findings reasonably satisfy criteria for the phenomenon of 'blind smell.'" In support of the previous findings of Sobel and his colleagues, they observe:

> An individual may be blind to (i.e., consciously unaware of) a smell, yet manifest reliable nonconscious responses to that smell. Of course, it is evident that Patient S. has lost more than mere olfactory awareness and exhibits only a rudimentary preservation of odor-detection ability. . . . However, the demonstration of odor-related activity in the left OFC . . . implies that the left olfactory pathway retains substantial functionality to implement olfactory sensory and affective analysis [that is unconscious to the subject].

In conclusion, this is a promising start to answering the question of where conscious perception of smell and flavor arise in the human brain flavor system, but we will need many more subjects and experiments like this to identify all the parts of the brain that contribute. With such a complex system, it is likely that conscious and unconscious perceptions will make their own contributions to the individual variations that make our flavor perceptions so fascinating.

CHAPTER TWENTY-SIX

Smell and Flavor in Human Evolution

The new evidence regarding how the brain creates smell and flavor that dominate our daily lives suggests the hypothesis that the human brain flavor system may have played a much larger role in human evolution than is appreciated. There has been little discussion of this possibility up to now. Speculations on how evolution occurred are notoriously difficult to make. However, in pursuing this question I have been greatly encouraged by interactions with anthropologists. It will be useful to identify some of the evidence for events in human evolution in which the human brain flavor system may have played a significant role. Five kinds of evidence seem most interesting: (1) the record of the genes; (2) competition between smell and sight; (3) the increase in brain size; (4) the adaptations of the musculoskeletal system for searching for food; and (5) the control of fire and the development of human cuisines.

The Record of the Genes

We have seen that the olfactory receptor genes are believed to be the largest family in the mammalian genome, accounting for 2 to 5 percent of the total, depending on one's estimate of the total number of genes. The large gene families were likely also characteristic of the ancient ancestors; dynamic changes presumably occurred constantly over time but the approximate sizes of the families were maintained because of their adaptive value.

Mammalian and primate evolution accelerated with the extinction of the dinosaurs around 65 million years ago. The original large size of the mammalian receptor family was more or less maintained as mammals diverged into their different orders, families, and species, as attested to by the large repertoire present in most mammals today. It is also likely that, as in modern mammals, the receptor cells projected to the glomerular sheet in the olfactory bulb, where spatial patterns were elicited by the different odors as we discussed in chapter 7. This assumption is strongly supported by the fact that glomeruli are a phylogenetic constant for the processing of smell across almost all vertebrates as well as invertebrates, as John Hildebrand and I reviewed in 1997. The olfactory system thus provides a backward reach into evolutionary time. It may be unique among brain systems for this purpose because of the close relation between the receptor genes and the images of the odor world that they form in the brain.

What was happening to the mammalian smell receptor repertoire during the evolution of modern primates, including humans? The genes for several species were sequenced and compared by Sylvie Rouquier and her colleagues in 2000. When arranged in order of branching from the primate tree, it appears that the number of functional genes has undergone a steady decline, from almost 100 percent in lemurs through the modern species of howler, macaque, baboon, chimp, to humans at about 35 percent. On the surface, this would seem to confirm the decline of the importance of smell in humans as opposed to most other mammals. However, we've seen that this conclusion has been challenged by Matthias Laska and his colleagues, then in Munich, who from testing monkeys and comparing with humans has provided evidence that primate olfaction is quite good compared with rodents and even dogs. Furthermore, in chapter 9 we saw that the number of receptors has to be evaluated within the context of the number of receptor cells and the number of olfactory glomeruli; the correlations appear to be complex. In addition, a major thesis of this book is that the declining numbers of olfactory receptor genes during human evolution are offset by brain size, as the forebrain expanded vigorously, and the olfactory pathway still had its privileged direct access. There were thus two balancing trends, a gradual reduction in the peripheral smell sensory receptor repertoire, offset by a large expansion of the central brain systems for analyzing the images. The expansion associated with consumption of more flavorful foods enhanced the system that created the flavors.

Competition Between Smell and Sight

The first primates of modern aspect appeared about 60 million years ago. These animals have dental and anatomical adaptations that indicate omnivory was a typical and successful primate adaptation. Their skeletal features suggest increases in manipulative abilities, a variety of locomotor patterns, and a reliance on vision. Because of forward-facing orbits and reduced snouts, lists of early primate features usually include reduced reliance on the olfactory sense. It is believed that the forward movement of the eyes required a reduction of the snout and, with it, a reduction in the size of the olfactory sensory receptor population and therefore in the smell part of the brain. This presumed decline of smell in the evolution of primates is one of the most entrenched ideas in anthropology and in the public mind.

The new evidence regarding the brain flavor system indicates that this idea needs to be revised. An increase in visual acuity does not necessarily mean reduced olfactory sensitivity and discrimination. A major new theme is that a smaller olfactory receptor repertoire in primates is offset by the larger brain size and the more complex processing of the smell signals. This reassessment is supported by behavioral studies that indicate that monkeys and humans have senses of smell that are quite good.

Like other mammals, therefore, primates likely were dominated by their sense of smell. The ripeness of fruit can be determined by feel or by sight, but perhaps most effectively by scent. Fruit eating in the jungle also required a more complex lifestyle: remembering when fruit is likely to ripen, making the necessary plans to be there at the right time, feeding the infants, and sharing among the members of the clan. More complex brains, dominated by an expanding neocortex, gave a competitive advantage in organizing these activities. In addition, like modern primates they likely used scent to identify individuals and to determine the social standing of particular animals.

Life among the trees thus required smell and larger brains and also required an excellent sense of vision. Most primates therefore have large eyes to enable both day and night vision. Many species also benefit from excellent color vision, using color to aid, for example, in judging the ripeness of fruit. High-acuity eyesight was also useful in plucking the fruit and bringing it close to the face for examining it visually for ripeness. In our view from the perspective of the nose, thus began one of the driving forces in primate evolution: the competition between smell and sight for control of the neocortex in primate behavior.

Increasing Brain Size

The earliest humans branched off from the lines leading to modern-day African apes about 5 to 6 million years ago. At that time, climate changes were causing a decrease in rainfall and the coming of different seasons. The environment changed, too. In Africa, the dense forests with their profusion of plant life and fruit trees were giving way to stands of trees scattered over grasslands and semideserts. It is hypothesized that in such a landscape, bipedalism gave early hominids an adaptive advantage in their search to find the increasingly scattered sources of food. Because these creatures were omnivorous, their diet probably included leafy greens, flowers, nuts, small vertebrates, insects, and fruits. One could imagine that because of their high sugar content, fruits were preferred food items for our hominid ancestors. As Rick Potts has noted, "The ephemeral aspect of high-quality fruit has placed a premium on cognitive and social means of finding and defending food sources." These cognitive and social demands are regarded as critical elements in the evolution of the primate brain leading to humans. From the present perspective, the importance of smell and flavor could be recognized in motivating the search, guiding it to its target fruit, discerning ripeness, and finding gratification in the consumption. In this view, it was not diet per se that was the initiating factor underlying the expansion of the primate brain, but the search for flavor. From the perspective of neuroeconomics (chapter 22), flavor put values on the food in the diet.

In the consensus view, early bipedal hominids were able to hunt more successfully for food, mainly fruit. In our view through the nose, they were motivated to find it by its smell, they judged its ripeness mainly by its smell, and they obtained pleasure and reward from eating it by its smell and flavor.

Achieving bipedalism required a number of delicate adjustments to the musculoskeletal system. It is tempting to suggest that many of them also served to acquire flavorful foods. Adaptations of the pelvis enabled a hominid to squat while grasping food with its hands, manipulating it, smelling it, and eating it. Keen eyesight enabled it to examine the food closely. An enlarging neocortex provided fine motor control of the fingers, which, together with high-acuity eyesight, enabled the hominid to search for other food sources. Adaptations of the motor pathways of the brain brought cortical control of each separate finger, so that individual seeds on the ground could be picked up and eaten. This fine control of the fingers occurred through direct connections from the motor nerve cells in the neocortex to the motor neurons in the spinal cord that innervate the finger muscles. A sharp sense of smell enabled these creatures not only to test every food by sniffing it but also to sense the flavors of ingested foods. Because the early hominids were omnivores, the foods consisted of novel mixtures, giving rise to new flavors from combinations of fruits, seeds, leaves, and small animals. It was the first step toward cuisine.

Around 2 million years ago a distinctly new type of hominid (*Homo ergaster* or *Homo erectus*) emerged, distinguished by a larger, more habitually upright body and a larger brain, both in absolute terms and relative to body size. It is assumed that these anatomical changes reflect a significant dietary change, including plenty of fat for the myelin that would have been needed in the larger brain. The diet of these humans, like that of modern-day humans, remained that of an omnivore, containing fruits and vegetables, roots, insects, and honey in addition to meat and marrow. It may be hypothesized that the flavors were learned to be advantageous for health, although the connection between flavor and nutrition remains one of the most challenging problems in research on feeding behavior.

Musculoskeletal Adaptations for Acquiring Flavorful Food

According to the current consensus, about 2 million years ago some humans migrated out of Africa, and some reached as far as Indonesia in a very short (evolutionarily speaking) period of time. This may have been facilitated by not only the larger upright stature and larger brains, but also the skeletal changes that appear to reflect adaptations for long-distance walking and running. What does this have to do with smell?

Among my anthropologist colleagues, Daniel Lieberman has shared much of my enthusiasm for rethinking the role of flavor in human evolution. We have seen that he has carried out deep studies of the evolution of the human head, providing new insights into the structural changes that would have contributed to retronasal smell. The adaptations of the head and body for long-distance running have been another area of interest. Dennis Bramble and Lieberman, in an article published in 2004, list 26 derived features of the skeleton that would have aided long-distance running, such as longer leg bones (for longer strides), wider joint cartilages (to absorb jolts), and more flexible relations between torso and limbs to allow better balance while running. These adaptations thus must be added to those for reaching, squatting, and examining food items that we have mentioned previously. It is striking that a short snout is listed as an adaptation for running and walking by making the head more balanced and thus aiding in head stabilization. This is a good example of how one has to take a broad behavioral perspective on the basis for biological adaptations. These and other adaptations are explained in full in Lieberman's book *The Evolution of the Human Head*.

What induced these early humans to journey thousands of miles over the Asian continent to China and Southeast Asia? Some cite wanderlust, a human curiosity to explore. But could these humans have been motivated to find new sources of foods? From the nose's perspective, a new possibility is the motivation to find plants that added flavors to the diet, the ancestral equivalents of herbs and spices. Later waves of human migrations over the spice trade routes during recorded history were driven by this craving for flavors. As we have seen, the craving center is a central component of the human flavor system, and smell is the main input.

Fire, Flavor, and Cuisine

A critical event was the use of cooking for preparing foods. When it began, and what has been its significance, has been controversial. The earliest statement I have found is from the late eighteenth century by

James Boswell, best known as the biographer of Samuel Johnson, but also a devoted gourmand:

> Dr. [Benjamin] Franklin said, Man was a 'tool-making animal,' which is very well; for no animal but man makes a thing, by means of which he can make another thing. But this applies to very few of the species. My definition of Man is, a 'Cooking Animal.' The beasts have memory, judgement, and all the faculties and passions of our mind, in a certain degree; but no beast is a cook.

In recent decades, many authors from many fields—anthropology, history, cooking, archaeology, sociology—have echoed this claim: "[C]ooking establishes the difference between animals and people" (Claude Lévi-Strauss); cooking defines "the human essence" (Michael Symons); cooking is an "index of the humanity of humankind" (Felipe Fernández-Armesto). But Richard Wrangham, from whose book *Catching Fire: How Cooking Made Us Human* these quotations are taken, points out that these claims are that "cooking had shaped us, but they did not say why or when or how."

To answer those questions, Wrangham has drawn on several decades of fieldwork and research to identify what fire actually would have done to influence the evolution of early humans. For this, he has essentially carried out the kind of analysis of what fire does to make meat and vegetables more easily chewed and flavorful, in parallel with the similar analyses that current food critics and chefs are carrying out on modern cuisines under the banner of molecular gastronomy and other movements. This has allowed him to focus on the key contribution that cooking made: the increased amount of energy, and the rapidity with which it could be acquired, from cooked foods. He provides evidence that the use of controlled fire for this purpose was contemporaneous with the spurt in brain size that characterized the passage from *Homo erectus* to *Homo sapiens*. This would mean that fire was controlled more than a million years before the best evidence for it has been found, but his arguments are persuasive.

I met Wrangham almost a decade ago, after I contacted him to discuss the early evidence for odor images and retronasal smell. He kindly invited me to give a seminar to the Department of Anthropology at Harvard, which stimulated a lively discussion and gave me encouragement toward developing the human brain flavor system hypothesis that adds to his hypothesis by making more explicit the brain mechanisms involved.

Cooking is important culturally because of the complex social organization that arises around it. Some individuals provide meat, others gather vegetables and other food items; there must be people who store it and guard it; the food must be prepared; preparation requires receptacles, whether in the ground or made into bowls and pots; there must be tools fashioned into utensils to cut and stir; and the family or local group must organize itself for consuming the food together around the cooking site. There is general agreement, as indicated in the earlier quotations, and as expounded many years ago by Peter Farb and George Armelagos in their book *Consuming Passions: The Anthropology of Eating*, that cooking was the activity that more than any other formed human society. Our new perspective extends that view: flavor was the glue that bonded societies together around the shared prepared meal.

To organize all these activities, exchange of information was essential. Some kind of language must have emerged to knit these complex emerging societies together. Steven Pinker in *The Language Instinct: How the Mind Creates Language* suggests that language arose in humans well before anatomically modern humans populated the Old World 40,000 years ago. Although there is no direct evidence for the first use of language, we may hypothesize that it arose in close association with the emergence of the divisions of labor and complex social interactions we have noted that are related to cooking. The more complex those relations became, the more extensive the vocabulary and complex the grammar.

This is where smell again comes into the picture. Although it is rarely mentioned in traditional accounts of the diets of our ancestors, the sensory qualities were likely to have been one of the main topics of conversation around the campfire as the food preparations proceeded and as the food was tasted and consumed. This would apply to the aroma before eating and even more to the flavor of the ingested food, which as we have noted is mostly due to smell. This ties smell, flavor, and language together in a way seldom recognized: the smells and flavors of cooking were likely a prime factor in the development of language. At first this seems counterintuitive because of the common belief that it is difficult to describe smells in words. But if smells and flavors are patterns that the brain creates, as we have shown, the patterns are part of the reason smells and flavors played the role they did.

In conclusion, the idea that the sense of smell has been reduced in the course of human evolution is one of the most entrenched ideas in anthropology and in the public mind. A broad range of research indicates that it is time to revise this view and incorporate odor images, the perception of flavor, and the extensive brain mechanisms they engage into an enlarged and enriched understanding of the biological basis of human behavior.

CHAPTER TWENTY-SEVEN

Why Flavor Matters

For many readers, the evidence presented in this book will be useful in giving personal insights into how "flavor" exists not in our food but in the way it is created by our brains. It should stimulate new insights for chefs, food critics, consumers of fast food, and families around the dinner table.

In addition to these personal rewards, knowledge of how the brain creates flavor has important implications for public policies on food and nutrition. A possible advantage of a new term like *neurogastronomy* is that it can help focus public policy more effectively in applying advances in brain science to issues related to food and flavor. These issues are with us from conception to old age. In Shakespeare's *As You Like It*, the philosopher Jacques describes the seven ages of man; here, we will describe the six ages of flavor.

Flavor Effects in the Womb

Both animal experiments and human studies show that the flavors of a mother's food are transmitted through the amniotic fluid in the womb to the fetus, and that after birth this transmission can influence later flavor preferences and dislikes. We learned about this in animals from Patricia Pedersen when she joined our lab in 1980. In her dissertation work with Elliott Blass at Princeton, she showed that an odorous substance injected into the amniotic fluid became a preferred cue for suckling by the rat pups after birth. Others showed that substances injected into the amniotic fluid or flavors of food eaten by the pregnant rat become associated with preferred or nonpreferred foods much longer after birth.

Similar experiments in humans showed that flavors of the mother's diet are transmitted to the amniotic fluid and the mother's milk, and that they influence the infant's preferences. Benoist Schaal, Luc Marlier, and Robert Soussignan in Nouzilly, France, reported in 2000 that infants whose mothers consumed anise (which has a licorice-like flavor) in their diet had a greater preference for anise when tested just after birth than those whose mothers' diets had not included it. The investigators used the same facial expressions for indicating preference and rejection as used to test for different tastes in the experiments of Jacob Steiner mentioned in chapter 13. Julie Mennella and Gary Beauchamp at the Monell Chemical Senses Center in Philadelphia have explored this question in depth. In a typical study reported in 2011, "infants whose mothers were randomly assigned to drink carrot juice during the last trimester of pregnancy enjoyed carrot-flavored cereals more than infants whose mothers did not. . . . Thus, like other mammals, prenatal experiences with food flavors transmitted from the mother's diet to amniotic fluid lead to greater acceptance and enjoyment of these foods during weaning." Other studies have used stronger foods, such as garlic, in the mother's diet to show its presence in the amniotic fluid and its influence on after-birth preference lasting into childhood.

As explained in chapter 13, the neural system for the basic hedonic responses to taste, in terms of attraction or repulsion, is in the brain stem and is active in the newborn. The learning of these preferences in utero and their emotional expression are therefore incorporated into this hardwired system.

Flavor and Infants

This evidence indicates that flavor preferences of the mother affect flavor preferences of the infant. This influence is also carried in the mother's breast milk. In furthering their study, Beauchamp and Mennella had mothers consume carrot juice while breastfeeding during their infants' first three months. When weaning later took place, the infants showed a preference for carrot juice, a result similar to that of the in utero experiments.

These authors have also found a sensitive learning period when infants can be trained to different flavors. They demonstrated this by feeding infants either standard formula or a protein-rich, less flavorful formula. Up to six months of age, infants learn to accept the latter, as judged by the facial expressions of acceptance; but their acceptance changes to rejection after six months, when the infants start showing a high preference for the particular taste on which they have been trained.

Beauchamp and Mennella conclude that "infants form a relatively precise flavor image which, when later matched, enhances intake and elicits pleasure." This *flavor image* for infants is the same term we have used for the adult. Because the infant brain lacks most of a functioning cerebral cortex, more research is needed to understand the extent of the developing human brain flavor system at these early stages of life.

Flavor in Childhood

There is widespread consensus that flavor has a powerful effect on young children. In fact, they are so vulnerable to flavors that Mennella has proposed that "children live in a different sensory world than adults." There is growing evidence that children prefer intense sweet, sour, and salty tastes and, if they are supertasters, are more sensitive to bitter tastes. This makes them especially vulnerable to the main culprits we have identified as leading to overeating—sweet foods, salt, and fat—through sensations that overwhelm the brain's control systems (see figure 22.1).

The nation's food producers and food providers of course understand this well. As noted in chapter 21, Eric Schlosser has documented how every item on fast-food menus is there to maximize the sensory stimulation and accompanying calorie overload. The fact that fast food appeals especially to kids means that with their preferences built into their flavor circuits at an early age by the plasticity we have discussed in chapter 23, costumers are guaranteed for life.

Perhaps no children's flavor item has been more targeted than sweet. This is a basic taste for the energy it provides, perhaps the most obvious and least sophisticated of the flavors. It is only as one grows up that the bitters—coffee, chocolate, beer—become favorites by their variety and deeper complexity. Kids are being overwhelmed by the sugar in soft drinks. When I was a kid, a standard Coke in a bottle or served by a soda jerk at the counter was 7 ounces (0.2 liter). Then Pepsi offered 12 ounces (0.35 liter, or, as the radio commercials sang, "twice as much for a nickel too"). Now the adult size of a soft drink is 16 or 20 ounces (around 0.5 or 0.6 liter), and 12 ounces is the children's size, almost twice the original adult size. Another example is dry cereal breakfast food. Nowadays it is difficult to find one that is not sugared. A blizzard of advertising—using cartoon characters, games, and television and on-line commercials—is aimed at children to eat them. In chapter 15, we noted that bright colors have a big effect on flavor, and this is especially so for children. Attempts are being made to regulate this advertising aimed at children by requiring the products to contain reduced amounts of sugar, salt, and fat.

More research is needed on children's brain flavor system in order to understand what makes it vulnerable. Using the control system in figure 22.1 as a guide, this research needs to test critical questions such as whether in children there is too high a sensitivity to the desired flavor image, too strong a memory of it, or an undeveloped prefrontal lobe system to control the motivation to acquire the desired foods. Research is also needed on the converse—that is, strong, hard-to-change aversions by some children to specific foods.

Flavor and Adolescents

The vulnerabilities of childhood continue into adolescence, magnified by new developments. Puberty brings a rush of hormones that loosen control as young people explore their newly developing worlds. Adolescents evolve from being largely dependent on their parents to having various degrees of partial and complete independence, gaining confidence to hang out with friends and make their own decisions. And, of course, they begin to have their own purchasing power to express their independence.

The maturing human brain flavor system reflects all these trends. Perhaps one of the most important factors in this is that the highest cognitive levels of the brain—the areas in the prefrontal cortex that are involved in making decisions on the basis of limited information, making plans with short-term as well as long-term consequences, and weighing immediate desires in the context of long-term goals—are still developing. This includes the areas of the human brain flavor system, which are still developing their key roles in creating flavor images and the images of desire for them. Little wonder that entry into the worldwide obesity epidemic begins in childhood and adolescence.

Advertisers for fast foods have known this continuing vulnerability for decades, and have targeted the baby-boomer youth culture to the greatest degree possible. As in the case of children, there needs to be a balance between the advertising and the nutrition of foods with desirable flavors. In response to Schlosser's book, most of the fast-food chains recalibrated their offerings. Regulations have evolved so that food packages in many Western countries contain data on calories, fat content, cholesterol, salt, and other ingredients, enabling even young people to begin to make informed choices. It was recently reported that a fast-food meal in Copenhagen contains just 0.012 ounce (0.35 gram) of trans fatty acids, whereas the same meal in New York contains more than 0.35 ounce (10 grams). There's still a long way to go.

Flavor and Diets

The hazards of too much flavor, of course, are endemic in the adult world. Flavor is much talked about but, as this book shows, inadequately understood. We are flooded with television programs about food, Web sites about food, and articles in the media about food. Good flavor is always the goal. Food critics provide many insights into food flavors. But how flavor is produced is usually in terms of what is in the food, rarely in terms of how the brain creates it. And so we continue to overeat and make ourselves vulnerable to obesity, diabetes, cardiovascular disease, stroke, and even certain cancers.

A key belief of most adults in controlling how much they eat is that they can do it through a diet. That is the basis, at least, for the ever-expanding diet industry. We are flooded with specialty diets, each carrying the name of its founder or its fad. Some focus on a basic nutrient—protein, carbohydrate, or fat—and either elevate or outlaw it. Others invoke a particular food group or national cuisine or a supposed early human diet. Key the word *diets* into Google and you can get a list of 89 or more. All have a claim on some sliver of the truth. And for most people, all fail.

This is a book about flavor, not diets, but it should be clear that the key element missing from most discussions of diet is flavor. A great deal of evidence has been presented here that has a direct bearing on why too many people eat too much. Sensory mechanisms are sensitive mainly to changes in their stimuli; this leads directly to the ability to be stimulated by new flavors even though one is overstimulated by previous flavors. There are strong ties to memory, making a flavor too vivid in one's thoughts. Favors are tied strongly to emotions, making it hard to resist a food with a favorite flavor. Research is increasingly uncovering evidence that strong desires for flavors and strong desires for drugs of abuse activate similar brain mechanisms. Many other examples of brain mechanisms in flavor that play a key role in overeating are found throughout this book.

It is my hope that increased research on these brain mechanisms can give us better insight into why we have difficulty in controlling what we eat. For example, when nutrition experts formulate official guidelines, there needs to be recognition that the human brain flavor system is going to be guiding consumers toward, or away from, the foods being recommended.

It is fortunate that awareness of the critical role of flavor is on the rise. In his book *Food Fight: The Inside Story of the Food Industry, America's Obesity Crisis, and What We Can Do About It*, Kelly Brownell, a colleague at Yale University and an authority on the obesity epidemic, cites Anthony Sclafani at Brooklyn College for his current studies of the overstimulation caused by the wide variety in high-calorie, highly flavorful foods produced by the food industry, and Barbara Rolls, now at Penn State University, for her studies of sensory-specific satiety that drives overeating these differently flavored foods, as discussed in chapter 21.

An important perspective on relating flavor to nutrition is provided by a story told by the well-known chef Jacques Pépin in 2006, reminiscing about his arrival in the United States in 1959. He describes how he had been working in a three-star restaurant in France but paradoxically sought his first job in the United States with Howard Johnson, one of the main American restaurant chains in the mid-twentieth century, before the rise of fast-food chains. Pépin was fascinated by how Howard Johnson's goal was to capture the essence of American meals, and he enthusiastically plunged into making such traditional American dishes as burgers, hot dogs, fried clams, macaroni and cheese, hash browns, ice cream sundaes, banana splits, apple pies, and specialty dishes such as beef burgundy stew, scallops in mushroom sauce, veal, and turkey—often prepared by the ton for thousands of patrons. He loved it!

The reason for mentioning this story is that Pépin emphasizes the essence of a traditional, pre-fast-food, American cuisine in its variety, flavor, and balanced nutritive values. Johnson and his chefs attempted to reproduce these qualities of the meal at home as closely as possible for the busy lunch-eater and the traveling motorist. They took great care in testing for flavor, using high-quality flavor ingredients, replacing margarine with butter, dehydrated onion with fresh onion, and frozen potatoes with real potatoes.

The menus contained a balance between the amount of the food and the flavor it contained. This meant that one became full from stretching the stomach in coordination with just the right amount of flavor before one's orbitofrontal cortex decided that one had enjoyed the pleasant flavor to satiety, before the food became aversive. Traditional cuisines have this balance; it is why they continue to be traditional. This balance is also the key to the principles of how the human brain flavor system should function.

Flavor and Old Age

In Shakespeare's play *As You Like It*, Jacques ends his soliloquy on the ages of man thusly:

> . . . Last scene of all,
> That ends this strange eventful history,
> Is second childishness and mere oblivion,
> Sans teeth, sans eyes, sans taste, sans everything.
>   *Act 2, scene 7*

And, he could have added, sans flavor.

In Shakespeare's day, with no real medicines and the average life span probably around 40 years, old age left most people in a ravaged state. In our day, there are many in their eighties and nineties still going strong. However, there are also many who are incapacitated or ill with one of the many infirmities of age. An overriding concern for their loved ones is a failure to thrive. This may have an organic cause, but in many cases its cause may be a loss of interest in food because of its lack of flavor. This afflicts many old people, both in hospitals and at home.

There is increasing interest in identifying these cases and treating them. In many cases we know the causes. On average, sensory abilities decline in later years. Some people fortunately may be little affected, but many suffer significant losses by natural ageing, quite apart from a disease process. Richard Doty of the Taste and Smell Center in Philadelphia has documented this with his Sniffin' Sticks tests and has shown a decline in smell sensitivity in the eighties and nineties. Diseases take their toll. It is now well documented that an early sign of Alzheimer's disease is a loss of smell, and the same occurs in other diseases such as Parkinson's. Given the key role of smell in flavor, it is therefore not surprising that many older and ill people lose their sense of flavor. We have noted in the introduction that a sudden loss of the sense of smell in younger adults can be devastating because of the loss of flavor, and there can be a similar effect in the aged. Failure to thrive can have many causes, but loss of flavor is one that is potentially treatable and should be checked first. Treating it depends on the person's natural preferences, but care has to be taken because some of the common ways of increasing flavor, such as adding salt, may be proscribed by the individual's medical condition.

Recent studies of Alzheimer's disease are revealing new and unexpected depths in its relation to the human brain flavor system. Jennifer Stamps and Linda Bartoshuk at the University of Florida have found, in studying flavor perception in a population of Alzheimer's patients, that some people who experienced a loss of taste also had a loss of retronasal smell for some foods, and that this loss reduced the flavor. Foods with strong associated touch sensations (nose-feel) were least affected. This suggested that it might be possible to enhance flavor in these subjects by specifically adding a "mouth-sense" stimulant such as pepper to the food. They found that adding the pepper to grape jelly as a test food gave an enhanced perception of the grape flavor because of the retronasal smell from the mouth, but it had no effect on the smell of the grape jelly when it was sniffed. This study thus adds to the ways that different senses can interact with one another to produce the kind of supra-additive effects discussed in chapter 14, and it indicates a strategy for helping with patients who are failing to thrive. It is also a reminder that one needs to test for multiple sensory losses in the aging and ill.

In summary, in treating failure to thrive, a good rule of thumb may be to reactivate the food cravings of childhood, enhancing the senses that contribute to flavor with strong smells, strong tastes, crunchy texture, bright colors, pleasant music—and talking pleasantly together as you eat the shared meal. Understanding the human brain flavor system can be just as important for the end of life as for the beginning.

Bibliography

Introduction

Gibbons, B. The intimate sense of smell. *National Geographic* 170 (1986): 324–361.

Gilbert, A. *What the Nose Knows: The Science of Scent in Everyday Life.* New York: Crown, 2008.

McGee, H. *On Food and Cooking: The Science and Lore of the Kitchen.* 2nd ed. New York: Scribner, 2004.

Shepherd, G. M. The human sense of smell: Are we better than we think? *PLoS Biology* 2 (2004): E146.

——. Smell images and the flavour system in the human brain. *Nature* 444 (2006): 316–321.

This, H. *Molecular Gastronomy: Exploring the Science of Flavor.* New York: Columbia University Press, 2006.

1. The Revolution in Smell and Flavor

Aristotle. *On the Soul. Parva Naturalia. On Breath* (ca. 350 B.C.E.). Translated by W. S. Hett. Loeb Classical Library 288. Cambridge, Mass.: Harvard University Press, 1957.

Blodgett, B. *Remembering Smell: A Memoir of Losing—and Discovering—the Primal Sense.* Boston: Houghton Mifflin Harcourt, 2010.

Brillat-Savarin, J. A. *The Physiology of Taste, or, Meditations on Transcendental Gastronomy* (1825). Translated by M. F. K. Fisher. Washington, D.C.: Counterpoint, 1999.

Finck, H. T. The gastronomic value of odours. *Contemporary Review* 50 (1886): 680–695.

Gilbert, A. *What the Nose Knows: The Science of Scent in Everyday Life*. New York: Crown, 2008.

Rozin, P. "Taste-smell confusions" and the duality of the olfactory sense. *Perception and Psychophysics* 31 (1982): 397–401.

Schlosser, E. *Fast Food Nation: The Dark Side of the All-American Meal*. New York: Perennial, 2002.

Thomas, L. *The Lives of a Cell: Notes of a Biology Watcher*. New York: Viking, 1974.

2. Dogs, Humans, and Retronasal Smell

Hepper, P. G., and D. L.Wells. How many footsteps do dogs need to determine the direction of an odour trail? *Chemical Senses* 31 (2006): 207–212.

Lieberman, D. E. *The Evolution of the Human Head*. Cambridge, Mass.: Harvard University Press, 2011.

Negus, V. E. *Comparative Anatomy and Physiology of the Nose and Paranasal Sinuses*. Edinburgh: Livingstone, 1958.

Settles, G. Sniffers: Fluid-dynamic sampling for olfactory trace detection in nature and homeland security. *Journal of Fluids Engineering* 127 (2005): 189–218.

Wrangham, R. *Catching Fire: How Cooking Made Us Human*. New York: Basic Books, 2009.

3. How the Mouth Fools the Brain

Cerf-Ducastel, B., and C. Murphy. fMRI activation in response to odorants delivered orally in aqueous solution. *Chemical Senses* 26 (2001): 625–637.

Veldhuizen, M. G., K. J. Rudenga, and D. M. Small. The pleasure of taste, flavor, and food. In *Pleasures of the Brain*, edited by M. L. Kringelbach and K. C. Berridge, 146–168. New York: Oxford University Press, 2010.

4. The Molecules of Flavor

Atkins, P. W. *Molecules*. New York: Freeman, 1987.

Brillat-Savarin, J. A. *The Physiology of Taste, or, Meditations on Transcendental Gastronomy* (1825). Translated by M. F. K. Fisher. Washington, D.C.: Counterpoint, 1999.

Carpino, S., S. Mallia, S. La Terra, C. Melilli, G. Licitra, T. E. Acree, D. M. Barbano, and P. J. Van Soest. Composition and aroma compounds of Ragusano cheese: Native pasture and total mixed rations. *Journal of Dairy Science* 87 (2004): 816–830.

Mayr, D., T. Mark, W. Lindinger, H. Brevard, and C. Yeretzian. In-vivo analysis of banana aroma by proteon transfer reaction–mass spectrometry. In *Flavour Research at the Dawn of the Twenty-first Century: Proceedings of the 10th Weurman Flavour Research Symposium*, edited by J.-L. Le Quéré and P. X. Etiévant, 256–259. London: Tec and Doc/Intercept, 2003.

McGee, H. *On Food and Cooking: The Science and Lore of the Kitchen.* 2nd ed. New York: Scribner, 2004.

Taylor, A. J. Volatile flavor release from foods during eating. *Critical Reviews in Food Science and Nutrition* 36 (1996): 765–784.

This, H. *Molecular Gastronomy: Exploring the Science of Flavor.* New York: Columbia University Press, 2006.

Wrangham, R. *Catching Fire: How Cooking Made Us Human.* New York: Basic Books, 2009.

5. Smell Receptors for Smell Molecules

Buck, L., and R. Axel. A novel multigene family may encode odorant receptors: A molecular basis for odor recognition. *Cell* 65 (1991): 175–187.

Lancet, D. Vertebrate olfactory reception. *Annual Review of Neuroscience* 9 (1986): 329–355.

Malnic, B., J. Hirono, T. Sato, and L. B. Buck. Combinatorial receptor codes for odors. *Cell* 96 (1999): 713–723.

Mori, K., and Y. Yoshihara. Molecular recognition and olfactory processing in the mammalian olfactory system. *Progress in Neurobiology* 45 (1995): 585–619.

Olfactory Receptor Database. *SenseLab.* Available at http://senselab.med.yale.edu/.

Singer, M. S. Analysis of the molecular basis for octanal interactions in the expressed rat 17 olfactory receptor. *Chemical Senses* 25 (2000): 155–165.

Zhao, H., L. Ivic, J. M. Otaki, M. Hashimoto, K. Micoshiba, and S. Firestein. Functional expression of a mammalian odorant receptor. *Science* 279 (1998): 237–242.

6. Forming a Sensory Image

Horseshoe crabs and vision. In *The Horseshoe Crab*. Available at http://www.mbl.edu/marine_org/images/animals/Limulus/vision/index.html.

Kuffler, S. W. Discharge patterns and functional organization of mammalian retina. *Journal of Neurophysiology* 16 (1953): 37–68.

Mach bands. Available at http://en.wikipedia.org/wiki/Mach_bands.

Mach, E. *The Analysis of Sensations* (1897). Translated by C. M. Williams and S. Waterlow. New York: Dover, 1959.

Ratliff, F. *Mach Bands: Quantitative Studies on Natural Networks in the Retina*. San Francisco: Holden Day, 1965.

Sterling, P., and J. B. Demb. Retina. In *The Synaptic Organization of the Brain*, edited by G. M. Shepherd, 217–270. 5th ed. New York: Oxford University Press, 2004.

7. Images of Smell

Adrian, E. D. Sensory messages and sensation: The response of the olfactory organ to different smells. *Acta Physiologica Scandinavica* 29 (1953): 5–14.

Leon, M., and B. Johnson. *Glomerular Activity Response Archive*. Available at http://gara.bio.uci.edu/.

SenseLab. Available at http://senselab.med.yale.edu/.

Sharp, F. R., J. S. Kauer, and G. M. Shepherd. Local sites of activity-related glucose metabolism in rat olfactory bulb during odor stimulation. *Brain Research* 98 (1975): 596–600.

Shepherd, G. M., W. R. Chen, and C. A. Greer. Olfactory bulb. In *The Synaptic Organization of the Brain*, edited by G. M. Shepherd, 165–216. 5th ed. New York: Oxford University Press, 2004.

Stewart, W. B., J. S. Kauer, and G. M. Shepherd. Functional organization of the rat olfactory bulb analyzed by the 2-deoxyglucose method. *Journal of Comparative Neurology* 185 (1975): 715–734.

Xu, F., C. A. Greer, and G. M. Shepherd. Odor maps in the olfactory bulb. *Journal of Comparative Neurology* 422 (2000): 489–495.

Youngentob, S. L., B. A. Johnson, M. Leon, P. R. Sheehe, and P. F. Kent. Predicting odorant quality perceptions from multidimensional scaling of olfactory bulb glomerular activity patterns. *Proceedings of the National Academy of Sciences of the United States of America* 104 (2007): 1953–1958.

8. A Smell Is Like a Face

Landau, T. *About Faces: The Evolution of the Human Face: What It Reveals, How It Deceives, What It Represents, and Why It Mirrors the Mind*. New York: Anchor, 1989.

Laska, M., D. Joshi, and G. M. Shepherd. Olfactory sensitivity for aliphatic aldehydes in CD-1 mice. *Behavioural Brain Research* 167 (2006): 349–354.

Matsumoto, H., K. Kobayakawa, R. Kobayakawa, T. Tashiro, K. Mori, H. Sakano, and K. Mori. Spatial arrangement of glomerular molecular-feature clusters in the odorant-receptor class domains of the mouse olfactory bulb. *Journal of Neurophysiology* 103 (2010): 3490–3500.

SenseLab. Available at http://senselab.med.yale.edu/.

Shepherd, G. M. Smell images and the flavour system in the human brain. *Nature* 444 (2006): 316–321.

Xu, F., N. Liu, I. Kida, D. L. Rothman, F. Hyder, and G. M. Shepherd. Odor maps of aldehydes and esters revealed by functional MRI in the glomerular layer of the mouse olfactory bulb. *Proceedings of the National Academy of Sciences of the United States of America* 100 (2003): 11029–11034.

9. Pointillist Images of Smell

Aungst, J. L., P. M. Heward, A. C. Puche, S.V. Karnup, A. Hayar, G. Szabo, and M. T. Shipley. Centre-surround inhibition among olfactory bulb glomeruli. *Nature* 426 (2003): 623–629.

Bozza, T., P. Feinstein, C. Zheng, and P. Mombaerts. Odorant receptor expression defines functional units in the mouse olfactory system. *Journal of Neuroscience* 22 (2002): 3033–3043.

Chen, W. R., and G. M. Shepherd. The olfactory glomerulus: A cortical module with specific functions. *Journal of Neurocytology* 34 (2006): 353–360.

Linster, C., and T. A. Cleland. Decorrelation of odor representations via spike timing-dependent plasticity. *Frontiers in Computational Neuroscience* 4 (2010): 157.

Maresh, A., D. Rodriguez Gil, M. C. Whitman, and C. A. Greer. Principles of glomerular organization in the human olfactory bulb—implications for odor processing. *PLoS One* 3 (2008): E2640.

Seurat, G. *A Sunday on La Grande Jatte* (1884). Available at http://www.artic.edu.

10. Enhancing the Image

Migliore, M., and G. M. Shepherd. Dendritic action potentials connect distributed dendrodendritic microcircuits. *Journal of Computational Neuroscience* 24 (2008): 207–221.

Pager, J. Ascending olfactory information and centrifugal influxes contributing to a nutritional modulation of the rat mitral cell responses. *Brain Research* 140 (1978): 251–269.

Rall, W., and G. M. Shepherd. Theoretical reconstruction of field potentials and dendro-dendritic synaptic interactions in olfactory bulb. *Journal of Neurophysiology* 31 (1968): 884–915.

Rall, W., G. M. Shepherd, T. S. Reese, and M. W. Brightman. Dendro-dendritic synaptic pathway for inhibition in the olfactory bulb. *Experimental Neurology* 14 (1966): 44–56.

Shepherd, G. M. Microcircuits in the nervous system. *Scientific American* 238 (1978): 92–103.

Shepherd, G. M., W. R. Chen, D. C. Willhite, M. Migliore, and C. A. Greer. The olfactory granule cell: From classical enigma to central role in olfactory processing. *Brain Research Reviews* 55 (2007): 373–382.

Willhite, D. C., K. T. Nguyen, A. V. Masurkar, C. A. Greer, G. M. Shepherd, and W. R. Chen. Viral tracing identifies distributed columnar organization in the olfactory bulb. *Proceedings of the National Academy of Sciences of the United States of America* 103 (2006): 12592–12597.

Xiong, W., and W. R. Chen. Dynamic gating of spike propagation in the mitral cell lateral dendrites. *Neuron* 34 (2002): 115–126.

11. Creating, Learning, and Remembering Smell

Douglas, R., K. A. C. Martin, and D. Whitteridge. A canonical microcircuit for neocortex. *Neural Compututation* 1 (1989): 480–488.

Gietzen, D. W., S. Hao, and T. G. Anthony. Mechanisms of food intake repression in indispensable amino acid deficiency. *Annual Review of Nutrition* 27 (2007): 63–78.

Haberly, L. B. Neuronal circuitry in olfactory cortex: Anatomy and functional implications. *Chemical Senses* 10 (1985): 219–238.

Haberly, L. B., and G. M. Shepherd. Current density analysis of summed evoked potentials in opposum prepyriform cortex. *Journal of Neurophysiology* 36 (1973): 789–802.

Hebb, D. O. *The Organization of Behavior*. New York: Wiley, 1949.

Neville, K. R., and L. B. Haberly. Olfactory cortex. In *The Synaptic Organization of the Brain*, edited by G. M. Shepherd, 415–454. 5th ed. New York: Oxford University Press, 2004.

Wilson, D. A., and R. J. Stevenson. *Learning to Smell: Olfactory Perception from Neurobiology to Behavior*. Baltimore: Johns Hopkins University Press, 2006.

12. Smell and Flavor

Kringelbach, M. L., and K. C. Berridge, eds. *Pleasures of the Brain*. New York: Oxford University Press, 2010.

Rolls, E. T. Taste, olfactory, and food texture reward processing in the brain and obesity. *International Journal of Obesity* 35 (2010): 550–561.

Schoenbaum, G., J. A. Gottfried, E. A. Murray, and S. J. Ramus, eds. *Linking Affect to Action: Critical Contributions of the Orbitofrontal Cortex*. Boston: Blackwell, for the New York Academy of Sciences, 2007.

Yeshurun, Y., and N. Sobel. An odor is not worth a thousand words: From multidimensional odors to unidimensional odor objects. *Annual Review of Psychology* 61 (2010): 219–241.

13. Taste and Flavor

Bartoshuk, L. Interview. *Science Careers* (blog), 2010. Available at http://blogs .sciencemag.org/sciencecareers/2010/06/linda-bartoshuk.html.

Brillat-Savarin, J. A. *The Physiology of Taste, or, Meditations on Transcendental Gastronomy* (1825). Translated by M. F. K. Fisher. Washington, D.C.: Counterpoint, 1999.

Brochet, F., and D. Dubourdieu. Wine descriptive language supports cognitive specificity of chemical senses. *Brain and Language* 77 (2001): 187–196.

Dinehart, M. E., J. E. Hayes, L. M. Bartoshuk, S. L. Lanier, and V. B. Duffy. Bitter taste markers explain variability in vegetable sweetness, bitterness, and intake. *Physiology and Behavior* 87 (2006): 304–313.

Garcia, J., and R. A. Koelling. Relation of cue to consequence in avoidance learning. *Psychonometric Science* 4 (1966): 123–124.

Small, D. M., J. C. Gerber, Y. E. Mak, and T. Hummel. Differential neural responses evoked by orthonasal versus retronasal odorant perception in humans. *Neuron* 47 (2005): 593–605.

Steiner, J. E. Discussion paper: Innate, discriminative human facial expressions to taste and smell stimulation. *Annals of the New York Academy of Sciences* 237 (1974): 229–233.

Verhagen, J. V., and L. Engelen. The neurocognitive bases of human multimodal food perception: Sensory integration. *Neuroscience and Biobehavioral Reviews* 30 (2006): 613–650.

Yarmolinsky, D. A., C. S. Zuker, and N. J. Ryba. Common sense about taste: From mammals to insects. *Cell* 139 (2009): 234–244.

14. Mouth-Sense and Flavor

Penfield, W., and T. Rasmussen. *The Cerebral Cortex of Man.* New York: Macmillan, 1950.

Verhagen, J. V., and L. Engelen. The neurocognitive bases of human multimodal food perception: Sensory integration. *Neuroscience and Biobehavioral Reviews* 30 (2006): 613–650.

Wang, G.-J., N. D. Volkow, C. Felder, J. S. Fowler, A. V. Levy, N. R. Pappas, C. T. Wong, W. Zju, and N. Netusil. Enhanced resting activity of the oral somatosensory cortex in obese subjects. *NeuroReport* 13 (2002): 1151–1155.

15. Seeing and Flavor

Brochet, F., and D. Dubourdieu. Wine descriptive language supports cognitive specificity of chemical senses. *Brain and Language* 77 (2001): 187–196.

Engen, T. The effect of expectation on judgments of odor. *Acta Psychologica* 36
(1972): 450–456.

Koza, B. J., A. Cilmi, M. Dolese, and D. A. Zellner. Color enhances orthonasal olfactory intensity and reduces retronasal olfactory intensity. *Chemical Senses* 30
(2005): 643–649.

Morrot, G., F. Brochet, and D. Dubourdieu. The color of odors. *Brain and Language*
79 (2001): 309–320.

Small, D. M., M. Jones-Gotman, R. Zatorre, M. Petrides, and A. Evans. Flavor processing: More than the sum of its parts. *NeuroReport* 8 (1997): 3913–3917.

16. Hearing and Flavor

Bourne, M. *Food Texture and Viscosity: Concept and Measurement*. 2nd ed. New
York: Academic Press, 2002.

Vickers, Z. M. Crispness and crunchiness—a difference in pitch? *Journal of Texture
Studies* 15 (1984): 157–163.

Vickers, Z. M., and M. C. Bourne. A psychoacoustical theory of crispness. *Journal of
Food Science* 41 (1976): 1158.

Vickers, Z. M., and S. S. Wasserman. Sensory qualities of food sounds based on individual perceptions. *Journal of Texture Studies* 10 (1980): 319.

17. The Muscles of Flavor

Amat, J.-M., and J.-D. Vincent. *L'art de parler la bouche pleine*. Paris: La Presqu'ile,
1997.

Brillat-Savarin, J. A. *The Physiology of Taste, or, Meditations on Transcendental
Gastronomy* (1825). Translated by M. F. K. Fisher. Washington, D.C.: Counterpoint, 1999.

Lieberman, D. E. *The Evolution of the Human Head*. Cambridge, Mass.: Harvard
University Press, 2011.

Linforth, R. S. T., A. Blissett, and A. J. Taylor. Differences in the effect of bolus
weight on flavor release into the breath between low-fat and high-fat products.
Journal of Agricultural and Food Chemistry 53 (2005): 7217–7221.

Matsuo, M., K. M. Hiiemae, M. Gonzalez-Fernandez, and J. B. Palmer. Respiration
during feeding on solid food: Alterations in breathing during mastication, pha-

ryngeal bolus aggregation, and swallowing. *Journal of Applied Physiology* 104 (2007): 674–681.

Pfeiffer, J. C., T. A. Hollowood, J. Hort, and A. J. Taylor. Temporal synchrony and integration of sub-threshold taste and smell signals. *Chemical Senses* 30 (2005): 539–545.

Tsachaki, M., R. S. T. Linforth, and A. J. Taylor. Aroma release from wines under dynamic conditions. *Journal of Agricultural and Food Chemistry* 57 (2009): 6976–6981.

18. Putting It Together

Bensafi, M., N. Sobel, and R. M. Khan. Hedonic-specific activity in piriform cortex mimics that during odor perception. *Journal of Neurophysiology* 98 (2007): 3254–3262.

Brillat-Savarin, J. A. *The Physiology of Taste, or, Meditations on Transcendental Gastronomy* (1825). Translated by M. F. K. Fisher. Washington, D.C.: Counterpoint, 1999.

Kosslyn, S. M. Mental images and the brain. *Cognitive Psychology* 22 (2005): 333–347.

Rinck, F., C. Rouby, and M. Bensafi. Which format for odor images? *Chemical Senses* 34 (2009): 11–13.

Veldhuizen, M. G., K. J. Rudenga, and D. M. Small. The pleasure of taste, flavor, and food. In *Pleasures of the Brain*, edited by M. L. Kringelbach and K. C. Berridge, 146–168. New York: Oxford University Press, 2010.

19. Flavor and Emotions

Franken, I. H. A., J. Booij, and W. van den Brink. The role of dopamine in human addiction: From reward to motivated attention. *European Journal of Pharmacology* 526 (2005): 199–206.

LaBar, K. S., A. C. Nobre, D. R. Gitelman, T. B. Parrish, Y.-H. Kim, and M.-M. Mesulam. Hunger selectively modulates corticolimbic activation to food stimuli in humans. *Behavioral Neuroscience* 113 (2001): 493–500.

Pelchat, M. L., A. Johnson, R. Chan, J. Valdez, and J. D. Ragland. Images of desire: Food-craving activation during fMRI. *NeuroImage* 23 (2004): 1486–1493.

Small, D. M., R. J. Zatorre, A. Dagher, A. C. Evans, and M. Jones-Gotman. Changes in brain activity related to eating chocolate: From pleasure to aversion. *Brain and Language* 124 (2001): 1720–1733.

20. Flavor and Memory

Fink, G. R., H. J. Markowitsch, M. Reinkemeier, T. Brudkbauer, J. Kessler, and W.-D. Heiss. Cerebral representation of one's own past: Neural networks involved in autobiographical memory. *Journal of Neuroscience* 16 (1996): 4275–4282.

Gottfried, J. A., A. P. R. Smith, M. D. Fugg, and R. J. Dolan. Remembrance of odors past: Human olfactory cortex in cross-modal recognition memory. *Neuron* 42 (2004): 687–695.

Lehrer, J. *Proust Was a Neuroscientist.* New York: Houghton Mifflin, 2007.

Proust, M. *Remembrance of Things Past.* Vol.1, *Swann's Way and Within a Budding Grove* (1913, 1919). Translated by C. K. Scott Moncrieff and T. Kilmartin. New York: Random House, 1981.

———. *Remembrance of Things Past.* Vol. 3, *The Captive, The Fugitive, Time Regained* (1923, 1925, 1927). Translated by C. K. Scott Moncrieff and T. Kilmartin. New York: Vintage, 1982.

Shepherd-Barr, K., and G. M. Shepherd. Madeleines and neuromodernism: Reassessing mechanisms of autobiographial memory in Proust. *Auto/Biography Studies* 1 (1998): 39–60.

21. Flavor and Obesity

Kessler, D. A. *The End of Overeating: Taking Control of the Insatiable American Appetite.* Emmaus, Pa.: Rodale, 2009.

McGraw, P. *The Ultimate Weight Solution Food Guide.* New York: Pocket Books, 2004.

Rolls, B. J., E. T. Rolls, E. A. Rowe, and K. Sweeney. Sensory specific satiety in man. *Physiology and Behavior* 27 (1981): 137–142.

Schlosser, E. *Fast Food Nation: The Dark Side of the All-American Meal.* New York: Perennial, 2002.

Small, D. M. Individual differences in the neurophysiology of reward and the obesity epidemic. *International Journal of Obesity* 33 Suppl. 2 (2009): S44–48.

Wang, G.-J., N. D. Volkow, C. Felder, J. S. Fowler, A. V. Levy, N. R. Pappas, C. T. Wong, W. Zju, and N. Netusil. Enhanced resting activity of the oral somatosensory cortex in obese subjects. *NeuroReport* 13 (2002): 1151–1155.

22. Decisions and the Neuroeconomics of Flavor and Nutrition

Corwin, R. L., and P. S. Grigson. Symposium overview—food addiction: Fact or fiction? *Journal of Nutrition* 139 (2009): 617–619.

Geisler, S., and D. S. Zahm. Afferents of the ventral tegmental area in the rat—anatomical substratum for integrative functions. *Journal of Comparative Neurology* 490 (2005): 270–294.

Glimcher, P. W. *Decisions, Uncertainty, and the Brain: The Science of Neuroeconomics.* Cambridge, Mass.: MIT Press, 2003.

Hare, T. A., C. G. Camerer, and A. Rangel. Self-control in decision-making involves modulation of the vmPFC valuation system. *Science* 324 (2009): 646–648.

Pelchat, M. L. Food addiction in humans. *Journal of Nutrition* 139 (2009): 620–622.

Schultz, W. Getting formal with dopamine and reward. *Neuron* 36 (2002): 241–263.

Shepherd, G. S. What can a research man say about values? *Journal of Farm Economics* 38 (1956): 8–16.

Volkow, N. D., G.-J. Wang, and R. D. Baler. Reward, dopamine and the control of food intake: Implications for obesity. *Trends in Cognitive Sciences* 15 (2011): 37–45.

23. Plasticity in the Human Brain Flavor System

Eliot, T. S. "The Dry Salvages." In *Four Quartets.* New York: Harcourt, Brace, 1943.

Gonzalez, K. M., C. Peo, T. Livdahl, and L. M. Kennedy. Experience-induced changes in sugar taste discrimination. *Chemical Senses* 33 (2008): 173–179.

Hebb, D. O. *The Organization of Behavior.* New York: Wiley, 1949.

Mainland, J. D., E. A. Brenner, N. Young, B. N. Johnson, R. M. Khan, M. Bensafi, and N. Sobel. Olfactory plasticity: One nostril knows what the other learns. *Nature* 419 (2002): 8902–8903.

Wang, H. W., C. J. Wysocki, and G. H. Gold. Induction of olfactory receptor sensitivity in mice. *Science* 260 (1993): 998–1000.

Wilson, D. A., and R. J. Stevenson. *Learning to Smell: Olfactory Perception from Neurobiology to Behavior.* Baltimore: Johns Hopkins University Press, 2006.

24. Smell, Flavor, and Language

Acree, T. E. *Flavornet,* 2003. Available at http://www.flavornet.org/flavornet.html.

Brochet, F., and D. Dubourdieu. Wine descriptive language supports cognitive specificity of chemical senses. *Brain and Language* 77 (2001): 187–196.

Burn, G. In pictures: Cy Twombly at Tate Modern. *Guardian,* June 14, 2008. Available at http://www.guardian.co.uk/books/2008/jun/14/saturdayreviewsfeatres .guardianreview17.

Miller, J. I. *The Spice Trade of the Roman Empire, 29 b.c.–a.d. 641.* New York: Oxford University Press, 1969.

Noble, A. *The Wine Aroma Wheel*, 2010. Available at http://winearomawheel.com/.

Parker, R. M., Jr. *Parker's Wine Buyer's Guide*. 4th ed. New York: Fireside, 1995.

Thomas, D. Chez Depardieu: The actor Gérard Depardieu's new restaurant is all the buzz in Paris. Dana Thomas books a table. *New York Times*, March 28, 2004, 18–22.

Walter, B. *Of Music and Music Making*. New York: Norton, 1961.

25. Smell, Flavor, and Consciousness

Crick, F., and C. Koch. A framework for consciousness. *Nature Neuroscience* 6 (2003): 119–126.

Li, W., I. Lopez, J. Osher, J. D. Howard, T. B. Parrish, and J. A. Gottfried. Right orbitofrontal cortex mediates conscious olfactory perception. *Psychological Science* 21 (2010): 1454–1463.

Sela, L., Y. Sacher, C. Serfaty, Y. Yeshurun, N. Soroker, and N. Sobel. Spared and impaired olfactory abilities after thalamic lesions. *Journal of Neuroscience* 29 (2009): 12059–12069.

Sobel, N., V. Prabhakaran, C. A. Hartley, J. E. Desmond, G. H. Glover, E. V. Sullivan, and J. D. E. Gabrieli. Blind smell: Brain activation induced by an undetected airborne chemical. *Brain* 122 (1999): 209–217.

Tait, D. S., and V. J. Brown. Difficulty overcoming learned non-reward during reversal learning in rats with Ibotenic acid lesions of orbital frontal cortex. In *Linking Affect to Action: Critical Contributions of the Orbitofrontal Cortex*, edited by G. Schoenbaum, J. A. Gottfried, E. A. Murray, and S. J. Ramus, 407–420. Boston: Blackwell, for the New York Academy of Sciences, 2007.

Verhagen, J. V. The neurocognitive bases of human multimodal food perception: Consciousness. *Brain Research Reviews* 53 (2007): 271–286.

26. Smell and Flavor in Human Evolution

Boswell, James. *Boswell's Life of Johnson, Together with Boswell's Journal of a Tour to the Hebrides and Johnson's Diary of a Journey into North Wales* (1785). 6 vols. Edited by G. B. Hill. Revised and enlarged by L. F. Powell. Oxford: Clarendon, 1934–1964.

Bramble, D. M., and D. E. Lieberman. Endurance running and the evolution of *Homo. Nature* 432 (2004): 345–352.

Farb, P., and G. Armelagos. *Consuming Passions: The Anthropology of Eating*. Boston: Houghton Mifflin, 1980.

Hildebrand, J. G., and G. M. Shepherd. Molecular mechanisms of olfactory discrimination: Converging evidence for common principles across phyla. *Annual Review of Neuroscience* 20 (1997): 595–631.

Laska, M., A. S. Seibt, and A. Weber. "Microsmatic" primates revisited: Olfactory sensitivity in the squirrel monkey. *Chemical Senses* 25 (2000): 47–53.

Lieberman, D. E. *The Evolution of the Human Head.* Cambridge, Mass.: Harvard University Press, 2011.

Pinker, S. *The Language Instinct: How the Mind Creates Language.* New York: Morrow, 1994.

Potts, R. Environmental hypotheses of hominim evolution. *Yearbook of Physical Anthropology* 41 (1998): 93–138.

Rouquier, S., A. Blancher, and D. Giorgi. The olfactory receptor gene repertoire in primates and mouse: Evidence for reduction of the functional fraction in primates. *Proceedings of the National Academy of Sciences of the United States of America* 97 (2000): 2870–2874.

Wrangham, R. *Catching Fire: How Cooking Made Us Human.* New York: Basic Books, 2009.

27. Why Flavor Matters

Beauchamp, G. K., and J. A. Mennella. Flavor perception in human infants: Development and functional significance. *Digestion* 83 Suppl. (2011): 1–6.

Brownell, K. D., and K. B. Horgren. *Food Fight: The Inside Story of the Food Industry, America's Obesity Crisis, and What We Can Do About It.* New York: McGraw-Hill, 2004.

Kristof, N. D. Killer girl scouts. *New York Times,* May 21, 2006, sec. 4, 15.

Mennella, J. A., C. P. Iagnow, and G. K. Beauchamp. Prenatal and postnatal flavor learning by human infants. *Pediatrics* 107 (2001): E188.

Pedersen, P. E., and E. M. Blass. Prenatal and postnatal determinants of the 1st suckling episode in albino rats. *Developmental Psychobiology* 15 (1982): 349–355.

Pépin, J. Howard Johnson's, adieu. *New York Times,* April 28, 2006, A23.

Schaal, B., L. Marlier, and R. Soussignan. Human foetuses learn odours from their pregnant mother's diet. *Chemical Senses* 25 (2000): 729–737.

Schlosser, E. *Fast Food Nation: The Dark Side of the All-American Meal.* New York: Perennial, 2002.

Stamps, J. F., and L. M. Bartoshuk. Rescuing flavor perception in the elderly. *AChemS Abstracts* P46 (2011).

Stickrod, G., D. P. Kimble, and W. P. Smotherman. In utero taste/odor aversion conditioning in the rat. *Physiology and Behavior* 28 (1982): 5–7.

Index